Calculus of Variations

Lev D. Elsgolc

Dover Publications, Inc.
Mineola, New York

Bibliographical Note

This Dover edition, first published in 2007, is an unabridged republication of the work published by Pergamon Press, New York, 1961. In this edition, two blank pages at the beginning of the original edition have been omitted, so that the present book begins on page 3.

International Standard Book Number
ISBN-13: 978-0-486-45799-4
ISBN-10: 0-486-45799-0

Manufactured in the United States by Courier Corporation
45799005   2015
www.doverpublications.com

CONTENTS

FROM THE PREFACE TO THE FIRST RUSSIAN EDITION

There has been in recent years a wide variety of applications of variational methods to various fields of mechanics and technology and this is why engineers, of many kinds, are faced with the necessity of learning the fundamentals of the calculus of variations.

The aim of this book is to provide engineers and students of colleges of technology with the opportunity of becoming familiar with the basic notions and standard methods of the calculus of variations including the direct methods of solution of the variational problems, which are important from the practical point of view.

Each chapter is illustrated by a large number of problems some of which are taken from existing textbooks.

INTRODUCTION

Along with the problems of discovering the maximum or minimum values of a given function $z = f(x)$, in engineering practice we often have to find the maxima or the minima of the values of mathematical entities called *functionals*.

Functionals are variable values which depend on a variable running through a set of functions, or on a finite number of such variables, and which are completely determined by a definite choice of these variable functions.

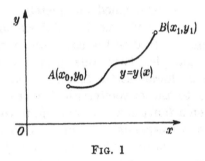

Fig. 1

For instance, the length l of a curve joining two given points on the plane is a functional, because this length is fully determined by choosing a definite function $y = y(x)$, the graph of which passes through these points (Fig. 1). As soon as the equation of the curve $y = y(x)$ is given, the value l can be calculated, namely,

$$l(y(x)) = \int_{x_0}^{x_1} \sqrt{1 + y'^2}\, dx.$$

Likewise, the area S of a surface is a functional. It is fully determined by choosing a definite surface, i.e. by choosing the function $z(x, y)$ that is involved in the equation of this surface $z = z(x, y)$. As is well known,

$$S\big(z(x, y)\big) = \iint_D \sqrt{1+\left(\frac{\partial z}{\partial x}\right)^2 +\left(\frac{\partial z}{\partial y}\right)^2}\, dx dy,$$

where D is the projection of this surface on the x, y-plane.

The moments of inertia of a homogeneous curve or a surface with respect to a point or an axis or a plane are also functionals. Their values are fully determined by choosing a curve or a surface, i.e. by choosing the functions that are involved in the equation of this curve or surface.

The resistance p, encountered by a physical body moving with given velocity in a medium, is also a functional. The value of p is determined completely by the shape of the surface of this body, i.e. it is determined by choosing the function involved in the equation of this surface.

All these examples have one property in common that is in fact a characteristic feature of all functionals, and analogous to the characteristic feature of ordinary functions. Given a functional $v = v\big(y(x)\big)$, to each function $y = y(x)$ there corresponds a unique number v, just as when we have an "ordinary" function $z = f(x)$, to each number x there corresponds a unique number z.

The variational calculus gives methods for finding the maximal and minimal values of functionals. Problems that consist in finding maxima or minima of a functional are called *variational problems*.

The variational calculus has been developing since 1696, and it became an independent mathematical discipline with its own research methods after the fundamental discoveries of a member of the Petersburg Academy of Sciences L. Euler (1707-1783), whom we can claim with good reason to be the founder of the calculus of variations.

The following three problems had considerable influence on the development of the calculus of variations.

The brachistochrone problem. In 1696 Johann Bernoulli published a paper, in which he suggested to mathematicians the problem of determining the path of quickest descent—the *brachistochrone*. This problem consists in finding which curve joining two given points A and B, not lying on the same vertical line, has the property that a massive particle sliding down along this curve from A to B reaches B in the shortest possible time (Fig. 2). It is easy to see that a path of quickest slide down is not a straight line joining the points A and B even though the straight line is the shortest line joining

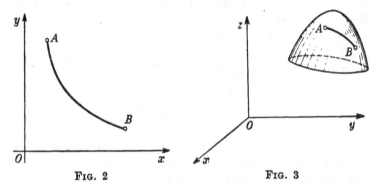

FIG. 2   FIG. 3

these points. When moving down a straight line a particle picks up speed comparatively slowly. If the line is steeper near the start point A, then its length will increase, but greater part of it will be done with more speed. The solution of the brachistochrone problem was given by Johann Bernoulli, Jacob Bernoulli, Newton and de l'Hospital. It turned out that the curve of quickest descent is a cycloid (see p. 38).

The problem of geodesics. This problem is to find the line of minimal length lying on a given surface $\varphi(x, y, z) = 0$ and joining two given points on this surface (Fig. 3). Such lines are called *geodesics*. This is a typical example of a variational problem of finding constrained extrema. We have to find the minimum of the functional

$$l = \int\limits_{x_0}^{x_1} \sqrt{1+y'^2+z'^2}\,dx,$$

where the functions $y(x)$ and $z(x)$ must satisfy the condition $\varphi(x, y, z) = 0$. This problem was solved in 1697 by Johann Bernoulli, but a general method of solving this type of problem was given by L. Euler and J. Lagrange.

   Isoperimetric problem. This problem is to find a closed curve of a given length l, encircling an area S that is maximal. It was known to the ancient Greeks, that such curve must be the circumference of a circle. The problem consists of finding the extrema of the functional S, under the additional condition that the length of the curve is constant, i.e. the functional

$$l = \int\limits_{t_0}^{t_1} \sqrt{x^{\cdot}(t)^2+y^{\cdot}(t)^2}\,dt \quad (*)$$

is kept constant. Such additional conditions are called *isoperimetric conditions*. The general methods of finding the solutions of problems with isoperimetric conditions were established in detail by L. Euler.

   We shall now show some methods of solving various variational problems.

   (*) Throughout this book the derivative with respect to t is denoted by $x^{\cdot}$.

CHAPTER I

THE METHOD OF VARIATION IN PROBLEMS WITH FIXED BOUNDARIES

1. The variation and its properties

The methods of solving variational problems, i.e. problems consisting of finding the maxima or minima of functionals, are very much like those of finding maxima or minima of ordinary functions. Therefore, it is not without interest to recall briefly the theory of maxima and minima of ordinary functions, and in parallel, to introduce the analogous notions for functionals and prove the analogous theorems.

1. The variable z is called a *function* of a variable x, in writing $z = f(x)$, if to each value of x from a certain domain there corresponds a certain value of z, i.e. to a given number x, there corresponds a number z.

1. The variable v is called a *functional* depending on a function $y(x)$, in writing $v = v(y(x))$, if to each function $y(x)$, from a certain class of functions, there corresponds a certain value of v, i.e. to a given function $y(x)$, there corresponds a number v.

2. The *increment* Δx of the argument x of a function $f(x)$ is the difference of two values of this argument $\Delta x = x - x_1$. If x is the independent variable, then the differential of x coincides with its increment $dx = \Delta x$.

2. The *increment* or the *variation* δy of the argument $y(x)$ of a functional $v(y(x))$ is the difference of two functions $\delta y = y(x) - y_1(x)$. It is assumed that the argument $y(x)$ runs through a certain class of functions.

3. A function $f(x)$ is said to be a *continuous function*, if small variations of x always lead to small variations of the function $f(x)$.

3. A functional $v(y(x))$ is said to be a *continuous functional*, if small variations of $y(x)$ always lead to small variations of the functional $v(y(x))$.

This latter definition can be made more precise and clearer. The question arises which variations of the function $y(x)$, that is itself an argument of a functional, are called small, or which curves $y = y(x)$, $y = y_1(x)$ are considered close to each other.

One possibility is to assume that the functions $y(x)$ and $y_1(x)$ are *close* to each other, whenever the absolute value of their difference, $y(x) - y_1(x)$ is small for all values of x for which the functions $y(x)$ and $y_1(x)$ are defined, i.e. such curves are considered close to each other that are coordinatewise close. Yet in many problems it is better to consider that only such curves are close that are not only coordinatewise close, but in addition, whose tangent lines at corresponding points have directions close to each other. That is, in such problems, the curves are called close, when not only the absolute value of the difference $y(x) - y_1(x)$ is small, but also the absolute value of the difference $y'(x) - y_1'(x)$.

It is sometimes necessary to assume that functions are close to each other only when the absolute value of all the differences

$$y(x) - y_1(x),\ y'(x) - y_1'(x),\ y''(x) - y_1''(x),\ \dots$$

$$\dots,\ y^{(k)}(x) - y_1^{(k)}(x),$$

are arbitrarily small.

Accordingly, we are lead to the following definition of closeness.

Two curves $y = y(x)$ and $y = y_1(x)$ are close or neighbouring in the sense of *closeness of order zero*, if the absolute value of the difference $y(x) - y_1(x)$ is small.

Two curves $y = y(x)$ and $y = y_1(x)$ are close or neighbouring in the sense of *closeness of order one*, if the absolute values of the differences $y(x) - y_1(x)$ and $y'(x) - y_1'(x)$ are small.

Two curves $y = y(x)$ and $y = y_1(x)$ are close or neighbouring in the sense of *closeness of order k*, if the absolute values of the differences

$$[y(x) - y_1(x), \; y'(x) - y_1'(x), \; \ldots, \; y^{(k)}(x) - y_1^{(k)}(x)$$

are small.

Two neighbouring curves in the sense of closeness of order zero, but not first-order close are shown at Fig. 4.

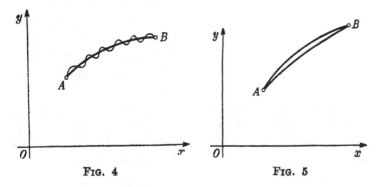

FIG. 4   FIG. 5

The coordinates of corresponding points are close, but the direction of tangent lines at such points varies considerably. An example of two first-order close curves is given at Fig. 5.

It follows directly from this definition that two neighbouring curves that are *k*-th order close, are also close in the sense of any order that is less than *k*.

The definition of continuity of a functional can now be stated more precisely as follows.

3'. A function $f(x)$ is *continuous at a point* $x = x_0$, if for each positive number ε there exists

3'. A functional $v(y(x))$ is *continuous along* $y = y_0(x)$ in the sense of closeness of order *k*, if for

a $\delta > 0$ such that

$$|f(x) - f(x_0)| < \varepsilon,$$

whenever

$$|x - x_0| < \delta.$$

It is also understood that x takes only such values for which the function $f(x)$ is defined.

arbitrary positive number ε there exists a $\delta > 0$ such that

$$\left| v\left(y(x)\right) - v\left(y_0(x)\right) \right| < \varepsilon,$$

whenever

$$|y(x) - y_0(x)| < \delta,$$

$$|y'(x) - y_0'(x)| < \delta,$$

$$\cdots\cdots\cdots\cdots$$

$$|y^{(k)}(x) - y_0^{(k)}(x)| < \delta.$$

It is understood, that the function $y(x)$ is taken from the class of functions for which $v\left(y(x)\right)$ is defined.

4. The function $l(x)$ is called *linear*, if the following conditions are satisfied:

$$l(cx) = cl(x),$$

where c is an arbitrary constant, and

$$l(x_1 + x_2) = l(x_1) + l(x_2).$$

Each linear function of one independent variable is of the form

$$l(x) = kx,$$

where k is a constant number.

4. The functional $L\left(y(x)\right)$ is called a *linear functional*, if it satisfies the following conditions:

$$L\left(cy(x)\right) = cL\left(y(x)\right),$$

where c is an arbitrary constant, and

$$L\left(y_1(x) + y_2(x)\right)$$
$$= L\left(y_1(x)\right) + L\left(y_2(x)\right).$$

For instance, the functional

$$L\left(y(x)\right)$$

$$= \int_{x_0}^{x_1} \left(p(x)y + q(x)y'\right)dx$$

is a linear functional.

5. If the increment

$$\Delta f = f(x + \Delta x) - f(x)$$

is of the form

$$\Delta f = A(x)\Delta x + \\ + \beta(x, \Delta x)\cdot\Delta x,$$

where $A(x)$ does not depend on Δx and $\beta(x, \Delta x) \to 0$ provided $\Delta x \to 0$, then $f(x)$ is called a *differentiable* function, and that part of the increment that is linear in Δx, namely $A(x)\Delta x$, is called the *differential* of $f(x)$ and is designated by df. Dividing this by Δx, and setting $\Delta x \to 0$ we obtain $A(x) = f'(x)$, and thence

$$df \doteq f'(x)\Delta x.$$

5. If the increment

$$\Delta v = v\big(y(x) + \delta y\big) - v\big(y(x)\big)$$

of a functional is of the form

$$\Delta v = L\big(y(x), \delta y\big) + \\ + \beta\big(y(x), \delta y\big)\max|\delta y|,$$

where $L\big(y(x), \delta y\big)$ is a linear functional in δy, $\max|\delta y|$ is the maximal value of $|\delta y|$, and $\beta\big(y(x), \delta y\big) \to 0$, whenever $\max|\delta y| \to 0$, then that part of this increment that is linear in δy, namely $L\big(y(x), \delta y\big)$ is called the *variation of the functional*, and it is designated by δv.

And so, the variation of a functional is in the main linear in the δy part of its increment.

The variation plays the same rôle in the theory of functionals as the differential does in that of ordinary functions.

It is also possible to define the differential of a function or the variation of a functional in another way which is almost equivalent. Let us consider the value of the function $f(x + \alpha\Delta x)$, where x and Δx are fixed and α is variable. Setting $\alpha = 1$ we obtain the increased value of the function $f(x + \Delta x)$, and setting $\alpha = 0$ we obtain the original value $f(x)$. It is easy to see that the derivative of $f(x + \alpha\Delta x)$ with respect to α at $\alpha = 0$, is the differential of $f(x)$ at the point x. In fact, according to the general rule for the differentiation of composite functions

$$\frac{\partial}{\partial a} f(x+a\,\varDelta x)|_{a=0} = f'(x+a\,\varDelta x)\,\varDelta x|_{a=0} = f'(x)\,\varDelta x = df(x).$$

The same holds of functions of several independent variables

$$z = f(x_1, x_2, \ldots, x_n).$$

The differential can be obtained by differentiating the function

$$f(x_1 + a\,\varDelta x_1, x_2 + a\varDelta x_2, \ldots, x_n + a\varDelta x_n)$$

with respect to a, and then setting $a = 0$. In fact,

$$\frac{\partial}{\partial a} f(x_1 + a\varDelta x_1, x_2 + a\,\varDelta x_2, \ldots, x_n + a\,\varDelta x_n)|_{a=0}$$

$$= \sum_{i=1}^{n} \frac{\partial f}{\partial x_i}\,\varDelta x_i = df.$$

In the same way, for functionals of the form $v\,(y(x))$ or even for more general ones, we can define the variation as the derivative of the functional $v\,(y(x) + a\delta y)$ with respect to a, at $a = 0$. In effect, if the variation of a functional in the sense of the main, linear part of its increment exists, then this increment has the form

$$\varDelta v = v\,(y(x) + a\,\delta y) - v\,(y(x))$$

$$= L(y,\, a\,\delta y) + \beta(y,\, a\,\delta y)|a|\max|\delta y|.$$

The derivative of $v(y + a\,\delta y)$ with respect to a, at $a = 0$ is

$$\lim_{\varDelta a \to 0} \frac{\varDelta v}{\varDelta a} = \lim_{a \to 0} \frac{\varDelta v}{a} = \lim_{a \to 0} \frac{L(y,\, a\,\delta y) + \beta(y(x),\, a\,\delta y)|a|\max|\delta y|}{a}$$

$$= \lim_{a \to 0} \frac{L(y,\, a\,\delta y)}{a} + \lim_{a \to 0} \frac{\beta(y(x),\, a\,\delta y)|a|\max|\delta y|}{a}$$

$$= L(y,\, \delta y),$$

for

$$L(y, a\,\delta y) = a\,L(y, \delta y)$$

and

$$\lim_{a\to 0} \frac{\beta(y(x), a\,\delta y)\,|a|\max|\delta y|}{a}$$

$$= \lim_{a\to 0} \beta(y(x), a\,\delta y)\max|\delta y| = 0,$$

as $\beta(y(x), a\,\delta y) \to 0$ whenever $a \to 0$.

And so, if the variation in the sense of the main linear increment of a functional exists, then the variation in the sense of the derivative with respect to a parameter at the initial value of this parameter exists too, and the two notions are equivalent.

6. The differential of a function $f(x)$ is given by	6. The variation of a functional $v(y(x))$ is given by
$$\frac{\partial}{\partial a}f(x + a\,\Delta x)\big|_{a=0}.$$	$$\frac{\partial}{\partial a}v(y(x) + a\,\delta y)\big|_{a=0}.$$

DEFINITION. A functional $v(y(x))$ takes on a maximum value along the curve $y = y_0(x)$, if all the values of this functional $v(y(x))$ taken on along arbitrary neighbouring to $y = y_0(x)$ curves are not greater than $v(y_0(x))$, i.e. $\Delta v = v(y(x)) - v(y_0(x)) \leqslant 0$. If $\Delta v \leqslant 0$ and $\Delta v = 0$ only when $y(x) = y_0(x)$, then we say that the functional $v(y(x))$ takes on an absolute maximum along the curve $y = y_0(x)$. Similarly we define a curve $y = y_0(x)$ along which the functional takes on a minimum value. In this case $\Delta v \geqslant 0$ for all curves sufficiently close to the curve $y = y_0(x)$.

7. THEOREM. *If a dif-*
ferentiable function $f(x)$
takes on a maximum or
minimum at an internal
point $x = x_0$, *then we have*

$$df = 0$$

at this point.

7. THEOREM. *If the*
variation of a functional
$v(y(x))$ *exists, and if* v
takes on a maximum or a
minimum along $y = y_0(x)$,
then

$$\delta v = 0$$

along $y = y_0(x)$.

Proof of the theorem for functionals: When $y_0(x)$
and δy are fixed, $v(y_0(x) + \alpha \delta y) = \varphi(\alpha)$ is a function of
α which takes on a maximum or a minimum for $\alpha = 0$.
Therefore

$$\varphi'(0) = 0 \ (^1)$$

or

$$\frac{\partial}{\partial \alpha} v(y_0(x) + \alpha \delta y)\big|_{\alpha=0} = 0,$$

i.e. $\delta v = 0$. And hence, the variation of a functional
vanishes along those curves which make the functional
an extremum.

The notion of extremum of a functional should be made
more precise. By a maximum or a minimum, or more
precisely, a relative maximum or minimum, we mean
the greatest or the smallest value of a functional among
those values that are taken on along neighbouring curves. However, as was shown above, the notion of closeness can be introduced in a variety of different ways, and
therefore when talking about maximum or minimum the
order of closeness should be pointed out distinctly.

If the curve $y = y_0(x)$ makes a functional $v(y(x))$
a maximum or minimum in the class of all curves for
which the absolute value of the difference $y(x) - y_0(x)$
is small, i.e. in the class of all neighbouring curves of

(^{1}) It is assumed that α can take on arbitrary values at a neighbourhood of the point $\alpha = 0$, positive, and negative as well.

$y = y_0(x)$ in the sense of closeness of order zero, then such maximum or minimum is called *strong*.

If the curve $y = y_0(x)$ makes a functional $v(y(x))$ a maximum or minimum, only in the class of neighbouring curves of $y = y_0(x)$ in the sense of first-order closeness, i. e. with respect to all curves that are not only coordinatewise close to $y = y_0(x)$, but coordinate and slopewise close to $y = y_0(x)$, then such maximum or minimum is called *weak*.

Obviously, if a curve $y = y_0(x)$ makes a functional a strong maximum (or minimum), then it makes this same functional a weak maximum (or minimum) too, since any neighbouring curve in the sense of closeness of the first order is also close in the sense of closeness of order zero. Yet it is possible that a weak maximum (or minimum) is taken on along a curve $y = y_0(x)$, but that it is not a strong maximum (or minimum). For, it may happen that there are no coordinate- and slope-wise close curves to $y = y_0(x)$ for which $v(y(x)) > v(y_0(x))$ (in case of a minimum $v(y(x)) < v(y_0(x))$), but among those curves which are only coordinatewise close we can find ones along which $v(y(x)) > v(y_0(x))$ (in case of a minimum $v(y(x)) < v(y_0(x))$). The distinction between strong and weak extrema does not make any difference to the proof of fundamental necessary condition for an extremum. This distinction however is essential when we investigate the sufficiency conditions for extrema (in Chapter III).

It should be observed that if a curve $y = y_0(x)$ makes $v(y(x))$ have an extremum, then not only

$$\frac{\partial}{\partial a} v(y_0(x) + a\,\delta y)\big|_{a=0} = 0,$$

but also

$$\frac{\partial}{\partial a} v(y(x, a))\big|_{a=0} = 0,$$

where $y(x, a)$ is an arbitrary class of admitted curves, such that, for $a = 0$ and $a = 1$, $y(x, a)$ becomes $y_0(x)$ or $y_0(x) + \delta y$ respectively. In fact, $v(y(x, a))$ is a function of a, for any specific choice of a determines a curve of the class $y(x, a)$ and hence a value of the functional $v(y(x, a))$.

Since $y_0(x)$ makes $v(y(x))$ an extremum, this function, too, has an extremum at $a = 0$, and therefore its derivative vanishes at $a = 0$[1].

Consequently

$$\frac{\partial}{\partial a} v(y(x, a))\Big|_{a=0} = 0,$$

but in general this derivative does not coincide with the variation of the functional, although as has just been shown it vanishes together with δv along the curves giving the extrema of the functional.

All definitions of this section as well as the fundamental theorem (p. 20) can be extended almost without change to functionals involving several independent functions

$$v(y_1(x), y_2(x), \ldots, y_n(x))$$

or one or more functions of several variables

$$v(z(x_1, x_2, \ldots, x_n)),$$

$$v(z_1(x_1, x_2, \ldots, x_n), z_2(x_1, x_2, \ldots, x_n), \ldots, z_m(x_1, x_2, \ldots, x_n)).$$

For instance, the variation δv of a functional $v(z(x, y))$ can be defined either as the main linear part of the increment

$$\Delta v = v(z(x, y) + \delta z) - v(z(x, y)),$$

---

[1] It is assumed that a takes on arbitrary values in a neighbourhood of the point $a = 0$, and $\dfrac{\partial v(y(x, a))}{\partial a}\Big|_{a=0}$ exists.

that is linear with respect to δz, or as the derivative with respect to a parameter at the initial value of the parameter

$$\frac{\partial}{\partial \alpha}\, v\big(z(x,\, y) + a\, \delta z\big)\big|_{\alpha=0}.$$

If $z = z(x,\, y)$ makes the functional v have an extremum, then $\delta v = 0$ along $z = z(x,\, y)$, for $v\big(z(x,\, y) + a\,\delta z\big)$ is a function of α which, according to the assumption just made, takes on an extremum at $\alpha = 0$ and therefore its derivative with respect to α at $\alpha = 0$ vanishes

$$\frac{\partial}{\partial \alpha}\, v\big(z(x,\, y) + a\, \delta z\big)\big|_{\alpha=0} = 0 \qquad \text{or} \qquad \delta v = 0.$$

2. Euler equation

Let us examine for extrema a functional of the simplest form

(1) $$v\big(y(x)\big) = \int_{x_0}^{x_1} F\big(x,\, y(x),\, y'(x)\big)\, dx,$$

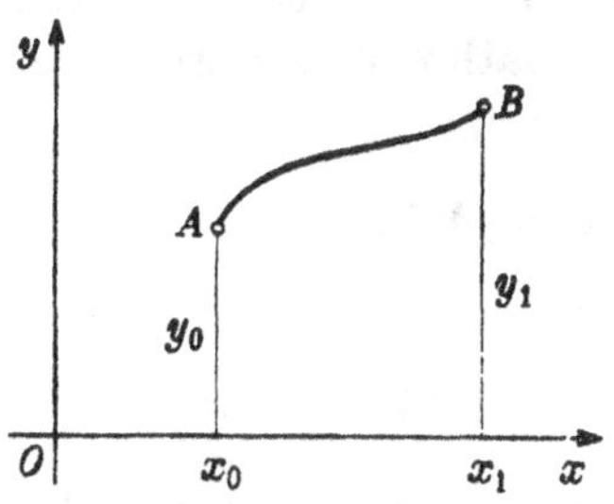

Fig. 6 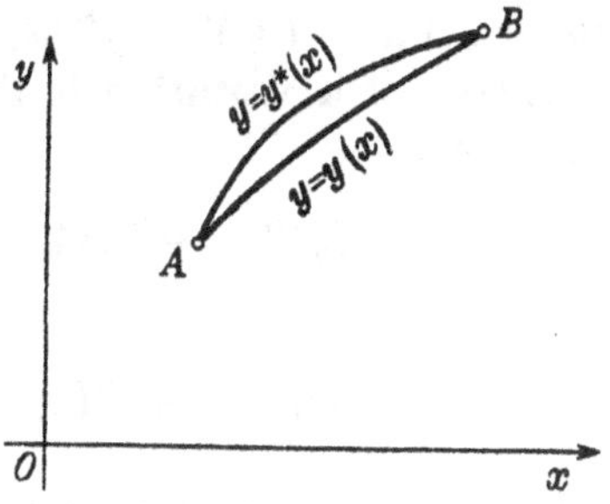   Fig. 7

where the end points $y(x_0) = y_0$ and $y(x_1) = y_1$ of admissible curves are fixed (Fig. 6). It is also assumed that the third derivative of the function $F(x, y, y')$ exists.

We already know that a necessary condition for an extremum of a functional is that its variation vanishes.

We are now going to show how this fundamental theorem applies to the functional in question. Let us assume that an extremum occurs along a curve $y = y(x)$ possessing a second-order derivative (supposing only that it has first-order derivative, we can prove by another method that the second-order derivative exists as well, provided the curve makes v an extremum). We take any admissible curve $y = y^*(x)$, neighbouring to $y = y(x)$, and we set up a one-parameter family of curves

$$y(x, a) = y(x) + a(y^*(x) - y(x))$$

containing the curves $y = y(x)$ and $y = y^*(x)$. When $a = 0$ we have $y = y(x)$, and when $a = 1$ we have $y = y^*(x)$ (Fig. 7). As we already know, the difference $y^*(x) - y(x)$ is called a *variation of the function* $y(x)$, and is designated by δy. The variation δy plays the same rôle in variational problems as the increment Δx of the independent variable does in the examination of an ordinary function $f(x)$ for extrema. The variation $\delta y = y^*(x) - y(x)$ is a function of x. This function can be differentiated once, or more, and we have $(\delta y)' = y^{*\prime}(x) - y'(x) = \delta y'$, that is, the derivative of a variation is the variation of a derivative, and likewise

$$(\delta y)'' = y^{*\prime\prime}(x) - y''(x) = \delta y'',$$
$$\cdot \cdot \cdot \cdot \cdot \cdot \cdot \cdot \cdot \cdot \cdot \cdot \cdot \cdot \cdot \cdot$$
$$(\delta y)^{(k)} = y^{*(k)}(x) - y^{(k)}(x) = \delta y^{(k)}.$$

Let us consider a family $y = y(x, a)$ or $y = y(x) + a\,\delta y$, that for $a = 0$ turns into a curve giving the functional an extremum, and for $a = 1$, into a certain neighbouring admissible curve, a so-called *comparison curve*.

If we consider the values of the functional

$$v(y(x)) = \int_{x_0}^{x_1} F(x, y, y')\,dx$$

taken on along the curves of the family $y = y(x, a)$ only, then we have a function of the variable a:

$$v\big(y(x, a)\big) = \varphi(a),$$

for in this instance the value $v\big(y(x, a)\big)$ of the functional depends only on a particular choice of the parameter a. Since for $a = 0$ we have $y = y(x)$, it follows that for $a = 0$ this function takes an extremum with respect to any neighbouring admissible curve and, in particular, with respect to any neighbouring curve of the family $y = y(x, a)$. As is well known, the necessary condition that the function $\varphi(a)$ has an extremum for $a = 0$ is that for $a = 0$ its derivative should vanish,

$$\varphi'(0) = 0.$$

Since

$$\varphi(a) = \int_{x_0}^{x_1} F\big(x, y(x, a), y_x'(x, a)\big)\, dx,$$

we have

$$\varphi'(a) = \int_{x_0}^{x_1} \left[F_y \frac{\partial}{\partial a} y(x, a) + F_{y'} \frac{\partial}{\partial a} y'(x, a) \right] dx,$$

where

$$F_y = \frac{\partial}{\partial y} F\big(x, y(x, a), y'(x, a)\big),$$

$$F_{y'} = \frac{\partial}{\partial y'} F\big(x, y(x, a), y'(x, a)\big).$$

Because of the relations

$$\frac{\partial}{\partial a} y(x, a) = \frac{\partial}{\partial a}\big(y(x) + a\delta y\big) = \delta y$$

and

$$\frac{\partial}{\partial a} y'(x, a) = \frac{\partial}{\partial a}\big(y'(x) + a\,\delta y'\big) = \delta y',$$

it follows that

$$\varphi'(a) = \int_{x_0}^{x_1} \left[F_y\big(x, y(x, a), y'(x, a)\big) \delta y + \right.$$

$$\left. + F_{y'}\big(x, y(x, a), y'(x, a)\big) \delta y' \right] dx,$$

$$\varphi'(0) = \int_{x_0}^{x_1} \left[F_y\big(x, y(x), y'(x)\big) \delta y + F_{y'}\big(x, y(x), y'(x)\big) \delta y' \right] dx.$$

As we have already remarked, $\varphi'(0)$ is called a variation of the functional and it is designated by δv. The necessary condition for a functional v to have an extremum is that its variation should vanish $\delta v = 0$.

In the case of the functional

$$v\big(y(x)\big) = \int_{x_0}^{x_1} F(x, y, y') \, dx$$

this condition is therefore

$$\int_{x_0}^{x_1} (F_y \, \delta y + F_{y'} \, \delta y') \, dx = 0.$$

Integrating by parts the second term, and remembering that $\delta y' = (\delta y)'$, we have

$$\delta v = [F_{y'} \, \delta y]_{x_0}^{x_1} + \int_{x_0}^{x_1} \left(F_y - \frac{d}{dx} F_{y'} \right) \delta y \, dx.$$

Since all the admissible curves pass through the fixed end points, it follows that

$$\delta y|_{x=x_0} = y^*(x_0) - y(x_0) = 0$$

and

$$\delta y|_{x=x_1} = y^*(x_1) - y(x_1) = 0,$$

and consequently

$$\delta v = \int_{x_0}^{x_1} \left(F_y - \frac{d}{dx} F_{y'} \right) \delta y \, dx.$$

Therefore, the necessary condition for an extremum takes the following form

(2)
$$\int_{x_0}^{x_1} \left(F_y - \frac{d}{dx} F_{y'} \right) \delta y \, dx = 0,$$

where the first factor $F_y - \dfrac{d}{dx} F_{y'}$ taken along the curve $y = y(x)$ that makes v an extremum is a certain continuous function, and the second factor δy may be an arbitrary function subject only to some general condition, for the comparison function $y = y^*(x)$ was chosen arbitrarily. This condition is that the function δy should vanish at the end points $x = x_0$ and $x = x_1$, should be a continuous function with first derivative or with derivatives of higher order, and either its absolute value $|\delta y|$ should be small, or both its absolute value $|\delta y|$ and the absolute value of its first derivative $|\delta y'|$ should be small.

To make the condition (2) simpler, the following lemma will be useful.

THE FUNDAMENTAL LEMMA OF THE CALCULUS OF VARIATIONS. *If a function $\Phi(x)$ is continuous in an interval (x_0, x_1) and if*

$$\int_{x_0}^{x_1} \Phi(x) \eta(x) \, dx = 0$$

for an arbitrary function $\eta(x)$ subject to some conditions of general character only, then $\Phi(x) \equiv 0$ throughout the interval $x_0 \leqslant x \leqslant x_1$. For instance the conditions that $\eta(x)$ should be a first or higher-order differentiable function, that $\eta(x)$ should vanish at the end points x_0, x_1, and $|\eta(x)| < \varepsilon$ or both $|\eta(x)| < \varepsilon$ and $|\eta'(x)| < \varepsilon$ are such general conditions.

Proof. We shall suppose that at a point $x = x^*$ of the interval $x_0 \leqslant x \leqslant x_1$, $\Phi(x^*) \neq 0$, and then show that this leads to a contradiction. In fact, since $\Phi(x)$ is continuous and $\Phi(x) \neq 0$, it follows that there is a neighbour-hood $x_0^* \leqslant x \leqslant x_1^*$ of the point x^* throughout which $\Phi(x)$ has a constant sign. Now, if we choose the function $\eta(x)$ so that it, too, has a constant sign in this neighbourhood and vanishes elsewhere (Fig. 8), then we have

$$\int_{x_0}^{x_1} \Phi(x)\eta(x)\,dx = \int_{x_0^*}^{x_1^*} \Phi(x)\eta(x)\,dx \neq 0,$$

for the product $\Phi(x)\eta(x)$ has a constant sign through-out the interval $x_0^* \leqslant x \leqslant x_1^*$ and vanishes outside. We

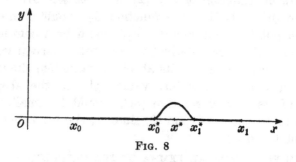

FIG. 8

have therefore obtained a contradiction, and hence $\Phi(x) \equiv 0$. For instance, we can choose the function $\eta(x)$ as follows:

$$\eta(x) = k(x - x_0^*)^{2n}(x - x_1^*)^{2n}$$

on the interval $x_0^* \leqslant x \leqslant x_1^*$,

where n is a positive integer and k is a constant number, and

$$\eta(x) = 0 \quad \text{elsewhere.}$$

Of course, this function satisfies the condition of the lemma, it has all continuous derivatives up to the order $2n-1$, and it vanishes at the points x_0 and x_1, and its absolute value as well as the absolute value of its deri-vatives can be made as small as necessary by choosing the coefficient k so that its absolute value is sufficiently small.

Remark. We can prove similarly that if the function $\Phi(x, y)$ is continuous in the domain D on the plane x, y, and $\iint_{D} \Phi(x, y) \eta(x, y) dx dy = 0$ for all arbitrarily chosen functions $\eta(x, y)$ satisfying only some conditions of general kind, then $\Phi(x, y) \equiv 0$ in the domain D. For instance, conditions of this kind are that $\eta(x, y)$ should be continuous, should have first derivatives or derivatives up to a certain higher order, should vanish at the boundary of D, and $|\eta| < \varepsilon$, $|\eta_x'| < \varepsilon$, $|\eta_y'| < \varepsilon$. For instance, in the proof of the fundamental lemma, we can define one of the functions $\eta(x, y)$ as follows:

$$\eta(x, y) = k\left((x-x^*)^2 + (y-y^*)^2 - \varepsilon_1^2\right)^{2n}$$

in the interior of a circle with the centre at a point (x^*, y^*) and a sufficiently small radius ε_1, the point (x^*, y^*) being so chosen that $\Phi(x^*, y^*) \neq 0$, and

$$\eta(x, y) \equiv 0$$

on the remaining part of D (Fig. 9).

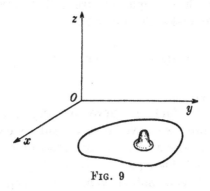

FIG. 9

We will now apply the fundamental lemma and simplify the condition

$$(2) \qquad \int_{x_0}^{x_1} \left(F_y - \frac{d}{dx} F_{y'}\right) \delta y \, dx = 0,$$

for an extremum of the most elementary form of a functional (1). All assumptions of the lemma are satisfied: the factor $F_y - \dfrac{d}{dx} F_{y'}$ taken along a curve $y = y(x)$ giving an extremum is a continuous function, and the variation δy is an arbitrary function with some restrictions of general kind that were considered in the lemma. Consequently, we have that along a curve that makes the functional v have an extremum $F_y - \dfrac{d}{dx} F_{y'} \equiv 0$, i.e. $y = y(x)$ is a solution of the second-order differential equation

$$F_y - \frac{d}{dx} F_{y'} = 0,$$

or written explicitly

$$F_y - F_{xy'} - F_{yy'} y' - F_{y'y'} y'' = 0.$$

This equation is called the *Euler equation*, after Euler who obtained it first in 1744. The integral curves $y = y(x, C_1, C_2)$ of the Euler equation are called *extremals*. Only extremals can make the functional

$$v(y(x)) = \int_{x_0}^{x_1} F(x, y, y') \, dx$$

have an extremum.

In order to find a curve making the functional (1) an extremum, we have to solve Euler's equation, and then determine the arbitrary constants involved in a general solution of this equation by making use of the conditions at the end points $y(x_0) = y_0, y(x_1) = y_1$. Only those extremals which satisfy these conditions can yield extrema. However, to make sure that they actually do give extrema, and to test which of them are maxima and which are minima, we must rely on the sufficient conditions for extrema, given in Chapter III.

EXAMPLE 1. *Which curves make the functional*

$$v(y(x)) = \int\limits_0^{\pi/2} ((y')^2 - y^2)\, dx, \quad y(0) = 0, \quad y(\tfrac{1}{2}\pi) = 1,$$

have an extremum?

The Euler equation is $y'' + y = 0$, with $y = C_1\cos x + C_2\sin x$ as a general solution. Making use of the conditions at the end points we have $C_1 = 0, C_2 = 1$. Consequently, only the curve $y = \sin x$ can give an extremum.

EXAMPLE 2. *Which curves can give an extremum of the functional*

$$v(y(x)) = \int\limits_0^1 ((y')_\frac{2}{4} + 12xy)\, dx, \quad y(0) = 0, \quad y(1) = 1?$$

The Euler equation is $y'' - 6x = 0$, and hence $y = x^3 + C_1 x + C_2$. From the boundary conditions we have $C_1 = 0, C_2 = 0$ and therefore an extremum can occur only along the curve $y = x^3$.

It was very easy to solve Euler equations in these two examples, but this is by no means the general situation, for second-order differential equations are integrable in terms of elementary functions only in exceptional cases. We now consider some simple instances in which Euler equations are integrable.

(1) F **does not depend on y':**

$$F = F(x, y).$$

The Euler equation is $F_y(x, y) = 0$, for $F_{y'} \equiv 0$. The solution of the equation $F_y(x, y) = 0$ involves no arbitrary constant, and therefore, in general, does not satisfy the boundary conditions $y(x_0) = y_0$ and $y(x_1) = y_1$.

Consequently, there is in general no solution of such a variational problem. Only in the exceptional case when the curve

$$F_y(x, y) = 0$$

passes through the end points (x_0, y_0) and (x_1, y_1) there exists a curve that can give an extremum.

EXAMPLE 3.

$$v(y(x)) = \int_{x_0}^{x_1} y^2\,dx, \qquad y(x_0) = y_0, \qquad y(x_1) = y_1.$$

The Euler equation is $F_y = 0$ or $y = 0$. The extremal $y = 0$ passes through the end points only for $y_0 = 0$ and $y_1 = 0$ (Fig. 10). Consequently, if $y_0 = 0$ and $y_1 = 0$, then the function $y = 0$ minimizes the functional $v = \int_{x_0}^{x_1} y^2\,dx$, for $v(y(x)) > 0$, and $v = 0$ when $y = 0$. Even if one of the values y_0 or y_1 is not zero, there is no continuous function minimizing this functional, which is rather evident, for we can choose a sequence of continuous functions whose graphs are going steeper and steeper from the point (x_0, y_0) down to the x-axis and then along the x-axis close to the point $(0, x_1)$, but again going upwards to (x_1, y_1) just before reaching the limit point x_1 (Fig. 11). It is evident that the values of the functional are arbitrarily small along the curves of such a sequence, and therefore the greatest lower bound of the functional is zero. However, this value cannot be attained along a continuous function, for along any continuous

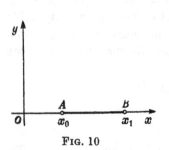

FIG. 10

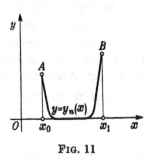

FIG. 11

function $y = y(x)$, not identically zero, the integral $\int_{x_0}^{x_1} y^2\,dx > 0$. This lowest bound of the value of the functional is attained along the curve with equations

$$y(x_0) = y_0,$$

$$y(x) = 0 \quad \text{for} \quad x_0 < x < x_1,$$

$$y(x_1) = y_1,$$

and this curve is not continuous (Fig. 12).

(2) The function F is linear in y':

$$F(x, y, y') = M(x, y) + N(x, y)y';$$

$$v(y(x)) = \int_{x_0}^{x_1} \left(M(x, y) + N(x, y) \frac{dy}{dx} \right) dx.$$

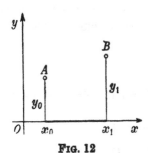

Fig. 12

The Euler equation is

$$\frac{\partial M}{\partial y} + \frac{\partial N}{\partial y} y' - \frac{d}{dx} N(x, y) = 0,$$

or

$$\frac{\partial M}{\partial y} + \frac{\partial N}{\partial y} y' - \frac{\partial N}{\partial x} - \frac{\partial N}{\partial y} y' = 0,$$

or

$$\frac{\partial M}{\partial y} - \frac{\partial N}{\partial x} = 0,$$

and again as in the previous case this is only a functional identity, not a differential equation. In general the curve

$$\frac{\partial M}{\partial y} - \frac{\partial N}{\partial x} = 0$$

does not satisfy the boundary condition, and consequently as a rule such a variational problem has no solution among continuous functions. If

$$\frac{\partial M}{\partial y} - \frac{\partial N}{\partial x} \equiv 0,$$

then $M\,dx + N\,dy$ is an exact differential and

$$v = \int_{x_0}^{x_1} \left(M + N\,\frac{dy}{dx} \right) dx = \int_{x_0}^{x_1} (M\,dx + N\,dy)$$

does not depend on the particular path of integration, and hence the functional has the same constant value along every admissible curve. The variational problem is of no interest.

EXAMPLE 4.

$$v(y\,(x)) = \int_0^1 (y^2 + x^2 y')\,dx, \quad y\,(0) = 0, \quad y\,(1) = a.$$

The Euler equation is

$$\frac{\partial M}{\partial y} - \frac{\partial N}{\partial x} = 0, \quad \text{or} \quad y - x = 0.$$

The first boundary condition $y\,(0) = 0$ is satisfied, but the second is satisfied only for $a = 1$. Consequently, if $a \neq 1$ there is no extremal satisfying these boundary conditions.

EXAMPLE 5.

$$v(y\,(x)) = \int_{x_0}^{x_1} (y + xy')\,dx, \quad \text{or} \quad v(y\,(x)) = \int_{x_0}^{x_1} (y\,dx + x\,dy),$$

$$y\,(x_0) = y_0, \quad y\,(x_1) = y_1.$$

The Euler equation is the identity $1 \equiv 1$. The integrand is an exact differential and hence the integral

$$v(y\,(x)) = \int_{x_0}^{x_1} d\,(xy) = x_1 y_1 - x_0 y_0,$$

does not depend on the curve of integration. The variational problem is of no interest.

(3) F depends on y' only:

$$F = F(y').$$

The Euler equation is $F_{y'y'}y'' = 0$, for $F_y = F_{xy'} = F_{yy'} = 0$. Consequently $y'' = 0$ or $F_{y'y'} = 0$. If $y'' = 0$, then $y = C_1 x + C_2$ is a two-parameter family of straight lines.

If the equation $F_{y'y'}(y') = 0$ has one or more real roots $y' = k_i$, then $y = k_i x + C$, and we have a one-parameter family of straight lines contained in the two-parameter family $y = C_1 x + C_2$ just derived. Therefore, in the case $F = F(y')$ the extremals are arbitrary straight lines $y = C_1 x + C_2$.

EXAMPLE 6. In case of the length of a path

$$l(y(x)) = \int_{x_0}^{x_1} \sqrt{1+y'^2}\, dx$$

the extremals are all straight lines $y = C_1 x + C_2$.

EXAMPLE 7. Let $t(y(x))$ be the time in which a particle moves from a point $A(x_0, y_0)$ to some other point $B(x_1, y_1)$ along a curve $y = y(x)$ with velocity $ds/dt = vy'$. If this velocity depends only on y', t is a functional of the form

$$t(y(x)) = \int_{x_0}^{x_1} \frac{\sqrt{1+y'^2}}{v(y')}\, dx,$$

$$\frac{ds}{dt} = v(y'), \quad dt = \frac{ds}{v(y')} = \frac{\sqrt{1+y'^2}\, dx}{v(y')}, \quad t = \int_{x_0}^{x_1} \frac{\sqrt{1+y'^2}}{v(y')}\, dx.$$

Consequently the extremals of this functional are straight lines.

(4) F depends only on x and y':

$$F = F(x, y').$$

The Euler equation

$$\frac{d}{dx} F_{y'}(x, y') = 0,$$

has an integral curve $F_{y'}(x, y') = C_1$, and as this first order equation $F_{y'}(x, y') = C_1$ does not involve y, it can be solved either by solving it for y' and integrating, or by introducing a suitably chosen parameter.

EXAMPLE 8. The functional

$$t(y(x)) = \int_{x_0}^{x_1} \frac{\sqrt{1+y'^2}}{x}\, dx$$

is the time that passes when a particle is moving from one point to some other point along the curve $y = y(x)$ with the velocity $v = x$, for if $ds/dt = x$, then $dt = ds/x$ and

$$t = \int_{x_0}^{x_1} \frac{\sqrt{1+y'^2}}{x}\, dx.$$

The first integral $F_{y'} = C_1$ of Euler's equation has the form $y'/x\sqrt{1+y'^2} = C_1$. The simplest method of solving this equation is by introducing a parameter, setting $y' = \tan t$. We have then

$$x = \frac{1}{C_1} \cdot \frac{y'}{\sqrt{1+y'^2}} = \frac{1}{C_1}\sin t,$$

or $x = C_1^* \sin t$, where $C_1^* = 1/C_1$;

$$\frac{dy}{dx} = \tan t, \quad dy = \tan t\, dx = \tan t \cdot C_1^* \cos t\, dt = C_1^* \sin t\, dt;$$

integrating, we obtain $y = -C_1^* \cos t + C_2$, and hence

$$x = C_1^* \sin t, \quad y - C_2 = -C_1^* \cos t$$

or, eliminating t, $x^2 + (y - C_2)^2 = C_1^{*2}$ and this is the family of all circles with centres at the y-axis.

(5) **F depends only on y and y':**

$$F = F(y, y').$$

The Euler equation is

$$F_y - F_{yy'} y' - F_{y'y'} y'' = 0,$$

for $F_{xy'} = 0$. If we multiply both sides by y', then the left-hand side of the Euler equation turns into an exact derivative $d(F - y' F_{y'})/dx$. In fact,

$$\frac{d}{dx}(F - y' F_{y'}) = F_y y' + F_{y'} y'' - y'' F_{y'} - F_{yy'} y'^2 -$$
$$- F_{y'y'} y' y'' = y'(F_y - F_{yy'} y' - F_{y'y'} y'').$$

Consequently, the Euler equation has a first integral

$$F - y' F_{y'} = C_1,$$

and since this first-order equation does not involve x explicitly, it can be solved by solving it for y' and by separation of variables or by introducing a parameter.

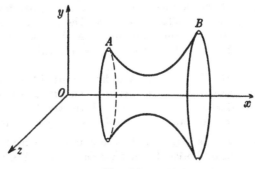

FIG. 13

EXAMPLE 9. The problem of minimum surface of revolution: *to determine a curve with given boundary points, such that by revolving this curve around the x-axis a surface of minimal area results* (Fig. 13).

As is well known the area of a surface of revolution is

$$S(y(x)) = 2\pi \int_{x_0}^{x_1} y\sqrt{1+y'^2}\,dx.$$

The integrand depends only on y and y', and therefore, the Euler equation has a first integral of the form

$$F - y'F_{y'} = C_1,$$

or in this particular case

$$y\sqrt{1+y'^2} - \frac{yy'^2}{\sqrt{1+y'^2}} = C_1.$$

The simplest way to find the general solution is to introduce a parameter $y' = \sinh t$, then $y = C_1 \cosh t$ and

$$dx = \frac{dy}{y'} = \frac{C_1 \sinh t\, dt}{\sinh t} = C_1\,dt, \quad x = C_1 t + C_2.$$

Hence the desired surface is obtained by revolving a curve with the parametric equations

$$x = C_1 t + C_2, \quad y = C_1 \cosh t.$$

Eliminating the parameter t, we have

$$y = C_1 \cosh \frac{x - C_2}{C_1},$$

which is a family of *catenaries*. A surface that is generated by rotation of a catenary is called a *catenoid*. The constants C_1 and C_2 are to be determined from the condition that the curve giving the solution of the problem should pass through the given end points.

EXAMPLE 10. The brachistochrone problem (see p. 11). *Find a curve joining two given points A and B so that a particle moving along this curve starting from A reaches B in the shortest time, friction and resistance of the medium being neglected.*

Let us place the origin of the coordinate system at the point A, the x-axis being horizontal and y-axis, vertical directed downwards. The velocity of the particle is $ds/dt = \sqrt{2gy}$, and consequently the time in which this point reaches $B(x_1, y_1)$ starting from $A(0, 0)$ is

$$t(y(x)) = \frac{1}{\sqrt{2g}} \int_0^{x_1} \frac{\sqrt{1 + y'^2}}{\sqrt{y}} \, dx, \quad y(0) = 0, \quad y(x_1) = y_1.$$

Since this functional is of the simplest form and the integrand does not depend explicitly on x, it follows that the Euler equation has a first integral $F - y'F_{y'} = C$, which in this particular case is

$$\frac{\sqrt{1 + y'^2}}{\sqrt{y}} - \frac{y'^2}{\sqrt{y(1 + y'^2)}} = C,$$

which reduces to $1/\sqrt{y(1 + y'^2)} = C$, or $y(1 + y'^2) = C_1$. Introducing a parameter, t, setting $y' = \cot t$, we obtain

$$y = \frac{C_1}{1 + \cot^2 t} = C_1 \sin^2 t = \frac{C_1}{2}(1 - \cos 2t),$$

$$dx = \frac{dy}{y'} = \frac{2C_1 \sin t \cos t \, dt}{\cot t} = 2C_1 \sin^2 t \, dt = C_1(1 - \cos 2t) \, dt,$$

$$x = C_1 \left(t - \frac{\sin 2t}{2} \right) + C_2 = \frac{C_1}{2}(2t - \sin 2t) + C_2.$$

Consequently, in parametric form, the equation of the curve is

$$x - C_2 = \frac{C_1}{2}(2t - \sin 2t), \quad y = \frac{C_1}{2}(1 - \cos 2t).$$

Now, if we observe that $C_2 = 0$, for $y = 0$ when $x = 0$, and make a modification of the parameter setting $2t = t_1$, we obtain a family of cycloids in the usual form

$$x = \frac{C_1}{2}(t_1 - \sin t_1), \quad y = \frac{C_1}{2}(1 - \cos t_1),$$

where $C_1/2$ is a radius of the rolling circle, which can be determined from the condition that the cycloid should pass through the point $B(x_1, y_1)$. Hence the brachistochrone has turned out to be a cycloid.

3. Functionals of the form:

$$\int_{x_0}^{x_1} F(x, y_1, y_2, \ldots, y_n, y_1', y_2', \ldots, y_n') dx$$

In order to obtain the necessary condition for a functional of more general form

$$v(y_1, y_2, \ldots, y_n) = \int_{x_0}^{x_1} F(x, y_1, y_2, \ldots, y_n, y_1', y_2', \ldots, y_n') dx$$

to have an extremum with prescribed boundary values for all the functions

$$y_1(x_0) = y_{10}, \quad y_2(x_0) = y_{20}, \quad \ldots, \quad y_n(x_0) = y_{n0},$$
$$y_1(x_1) = y_{11}, \quad y_2(x_1) = y_{21}, \quad \ldots, \quad y_n(x_1) = y_{n1},$$

we vary only one of the functions

$$y_j(x) \quad (j = 1, 2, \ldots, n),$$

at a time, keeping all the other functions fixed. By so doing the functional $v(y_1, y_2, \ldots, y_n)$ turns into a functional depending only on one function that is being varied, for instance on $y_i(x)$,

$$v(y_1, y_2, \ldots, y_n) = \tilde{v}(y_i).$$

The functional of this type was considered in § 2, and consequently a function that makes it have an extremum must satisfy the Euler equation

$$F_{y_i} - \frac{d}{dx} F_{y_i'} = 0.$$

Since this argument applies to each function y_i, $i = 1, 2, \ldots, n$, we have a system of second-order differential equations

$$F_{y_i} - \frac{d}{dx} F_{y_i'} = 0, \quad i = 1, 2, \ldots, n$$

which, in general, determines a $2n$-parameter family of integral curves in the space $x, y_1, y_2, \ldots, y_n$—the family of extremals for a given variational problem.

In particular, if the functional depends only on two functions $y(x)$, $z(x)$,

$$v\big(y(x), z(x)\big) = \int_{x_0}^{x_1} F(x, y, z, y', z')\, dx,$$

$$y(x_0) = y_0, \quad z(x_0) = z_0, \quad y(x_1) = y_1, \quad z(x_1) = z_1,$$

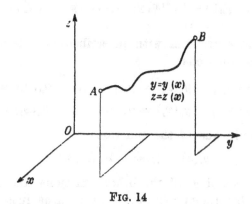

FIG. 14

i.e. any space curve $y = y(x)$, $z = z(x)$ (Fig. 14) determines a value of this functional, then varying only $y(x)$, $z(x)$ being fixed, we are varying the curve in such a way that its projection on the x, z-plane remains fixed, and consequently the curve itself remains on the cylinder $z = z(x)$ throughout the entire process of variation (Fig. 15).

Likewise, fixing $y(x)$ and varying $z(x)$, we are varying the curve in the way that it keeps lying on the cylinder

$y = y(x)$ throughout the process of variation. So doing, we obtain a system of two Euler equations

$$F_y - \frac{d}{dx} F_{y'} = 0 \quad \text{and} \quad F_z - \frac{d}{dx} F_{z'} = 0.$$

EXAMPLE 1. *Find the extremals of the functional*

$$v(y(x), z(x)) = \int_0^{\pi/2} (y'^2 + z'^2 + 2yz)\, dx,$$

$$y(0) = 0, \quad y(\pi/2) = 1, \quad z(0) = 0, \quad z(\pi/2) = -1.$$

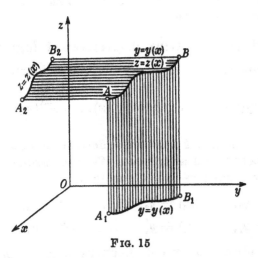

FIG. 15

The system of Euler equations is

$$y'' - z = 0, \quad z'' - y = 0.$$

Eliminating one of the unknown functions, for instance z, we have $y^{\mathrm{iv}} - y = 0$. This is a linear equation with constant coefficients whose solution is

$$y = C_1 e^x + C_2 e^{-x} + C_3 \cos x + C_4 \sin x,$$
$$z = y'' = C_1 e^x + C_2 e^{-x} - C_3 \cos x - C_4 \sin x.$$

From the boundary conditions, we have

$$C_1 = 0, \quad C_2 = 0, \quad C_3 = 0, \quad C_4 = 1,$$

and consequently,

$$y = \sin x, \quad z = -\sin x.$$

EXAMPLE 2. *Find the extremals of the functional*

$$v(y(x), z(x)) = \int_{x_0}^{x_1} F(y', z')\, dx.$$

The system of Euler equations is

$$F_{y'y'} y'' + F_{y'z'} z'' = 0, \quad F_{y'z'} y'' + F_{z'z'} z'' = 0$$

hence, supposing $F_{y'y'} F_{z'z'} - (F_{y'z'})^2 \neq 0$ we have: $y'' = 0$ and $z'' = 0$, or $y = C_1 x + C_2$, $z = C_3 x + C_4$, which is the family of straight lines in the space.

4. Functionals involving derivatives of higher order

We now examine the extrema of the functional

$$v(y(x)) = \int_{x_0}^{x_1} F(x, y(x), y'(x), \ldots, y^{(n)}(x))\, dx,$$

where the function F has partial derivatives of order $n+2$ with respect to all arguments. We also assume that the boundary conditions are of the form

$$y(x_0) = y_0, \quad y'(x_0) = y_0', \quad \ldots, \quad y^{(n-1)}(x_0) = y_0^{(n-1)},$$
$$y(x_1) = y_1, \quad y'(x_1) = y_1', \quad \ldots, \quad y^{(n-1)}(x_1) = y_1^{(n-1)},$$

i.e. not only the admitted functions take on prescribed values at the end points, but also their derivatives up to the order $n-1$ inclusively. Suppose that a $2n$ times differentiable curve $y = y(x)$ gives an extremum, and suppose $y = y^*(x)$ is a certain comparison curve, which is also $2n$ times differentiable.

We consider a one-parameter family of functions

$$y(x, a) = y(x) + a(y^*(x) - y(x)) \quad \text{or}$$
$$y(x, a) = y(x) + a\, \delta y.$$

For $a = 0$, $y(x, a) = y(x)$, and for $a = 1$, $y(x, a) = y^*(x)$. If we consider the values taken by the functional $v(y(x))$

along the curves of the family $y = y(x, a)$ only, then this functional turns into an ordinary function of the parameter a, that has an extremum for $a = 0$. Consequently,

$$\frac{d}{da} v\big(y(x, a)\big)\big|_{a=0} = 0.$$

As in Section 1 this derivative is called a variation of the functional v and it is designated by δv,

$$\delta v = \left[\frac{d}{da} \int_{x_0}^{x_1} F\big(x, y(x, a), y'(x, a), \ldots, y^{(n)}(x, a)\big) dx \right]_{a=0}$$

$$= \int_{x_0}^{x_1} (F_y \, \delta y + F_{y'} \, \delta y' + F_{y''} \, dy'' + \ldots + F_{y^{(n)}} \, \delta y^{(n)}) \, dx.$$

Integrating the second term of the right-hand side by parts, we have

$$\int_{x_0}^{x_1} F_{y'} \, \delta y' \, dx = [F_{y'} \, \delta y]_{x_0}^{x_1} - \int_{x_0}^{x_1} \frac{d}{dx} F_{y'} \, \delta y \, dx.$$

Similarly, integrating the third term by parts twice, we have

$$\int_{x_0}^{x_1} F_{y''} \, \delta y'' \, dx$$

$$= [F_{y''} \, \delta y']_{x_0}^{x_1} - \left[\frac{d}{dx} F_{y''} \, \delta y \right]_{x_0}^{x_1} + \int_{x_0}^{x_1} \frac{d^2}{dx^2} F_{y''} \, \delta y \, dx,$$

and so on, the last term being integrated by parts n times

$$\int_{x_0}^{x_1} F_{y^{(n)}} \, \delta y^{(n)} \, dx = [F_{y^{(n)}} \, \delta y^{(n-1)}]_{x_0}^{x_1} -$$

$$- \left[\frac{d}{dx} F_{y^{(n)}} \, \delta y^{(n-2)} \right]_{x_0}^{x_1} + \ldots + (-1)^n \int_{x_0}^{x_1} \frac{d^n}{dx^n} F_{y^{(n)}} \, \delta y \, dx.$$

Remembering that the end points $x = x_0$, $y = y_0$ are fixed, and consequently that all the variations taken at these points vanish, $\delta y = \delta y' = \delta y'' = \ldots = \delta y^{(n-1)} = 0$, we finally have

$$\delta v = \int_{x_0}^{x_1} \left(F_y - \frac{d}{dx} F_{y'} + \frac{d^2}{dx^2} F_{y''} - \frac{d^3}{dx^3} F_{y'''} + \ldots + \right.$$
$$\left. + (-1)^n \frac{d^n}{dx^n} F_{y^{(n)}} \right) \delta y \, dx.$$

Since along the curve giving an extremum we have

$$\delta v = \int_{x_0}^{x_1} \left(F_y - \frac{d}{dx} F_{y'} + \frac{d^2}{dx^2} F_{y''} + \ldots + \right.$$
$$\left. + (-1)^n \frac{d^n}{dx^n} F_{y^{(n)}} \right) \delta y \, dx = 0.$$

with arbitrary δy, and since the first factor of the integrand taken along the curve $y = y(x)$ is a continuous function of x, it follows from the fundamental lemma that this first factor is identically zero,

$$F_y - \frac{d}{dx} F_{y'} + \frac{d^2}{dx^2} F_{y''} + \ldots + (-1)^n \frac{d^n}{dx^n} F_{y^{(n)}} \equiv 0.$$

Consequently, any function that gives an extremum of the functional

$$v(y(x)) = \int_{x_0}^{x_1} F(x, y, y', y'', \ldots, y^{(n)}) \, dx,$$

must satisfy the equation

$$F_y - \frac{d}{dx} F_{y'} + \frac{d^2}{dx^2} F_{y''} + \ldots + (-1)^{(n)} \frac{d^n}{dx^n} F_{y^{(n)}} = 0.$$

This differential equation of order $2n$ is called the *Euler-Poisson equation*, and the integral curves of this equations are called *extremals* of the variational problem. The general solution of this equation involves $2n$ arbitrary constants, which can be determined, as a rule, from the $2n$ boundary conditions,

$$y(x_0) = y_0, \quad y'(x_0) = y_0', \quad \dots, \quad y^{(n-1)}(x_0) = y_0^{(n-1)},$$
$$y(x_1) = y_1, \quad y'(x_1) = y_1', \quad \dots, \quad y^{(n-1)}(x_1) = y_1^{(n-1)}.$$

EXAMPLE 1. *Find the extremal of the functional*

$$v(y(x)) = \int_0^1 (1 + y''^2)\, dx,$$

satisfying the boundary conditions

$$y(0) = 0, \quad y'(0) = 1, \quad y(1) = 1, \quad y'(1) = 1.$$

The Euler-Poisson equation is

$$\frac{d^2}{dx^2}(2y'') = 0$$

or $y^{\mathrm{iv}} = 0$. The general solution is

$$y = C_1 x^3 + C_2 x^2 + C_3 x + C_4.$$

From boundary conditions we have

$$C_1 = 0, \quad C_2 = 0, \quad C_3 = 1, \quad C_4 = 0.$$

Hence an extremum can be taken on only along the straight line $y = x$.

EXAMPLE 2. *Find the extremal of the functional*

$$v(y(x)) = \int_0^{\pi/2} (y''^2 - y^2 + x^2)\, dx.$$

satisfying the boundary conditions

$$y(0) = 1, \quad y'(0) = 0, \quad y(\pi/2) = 0, \quad y'(\pi/2) = -1.$$

The Euler-Poisson equation is $y^{\mathrm{iv}} - y = 0$. The general solution of this equation is $y = C_1 e^x + C_2 e^{-x} + C_3 \cos x + C_4 \sin x$. From boundary conditions we have $C_1 = 0$, $C_2 = 0$, $C_3 = 1$, $C_4 = 0$. Hence an extremum can be taken on only along the curve $y = \cos x$.

EXAMPLE 3. *Find the extremal of the functional*

$$v(y(x)) = \int_{-l}^{l} (\tfrac{1}{2}\mu y''^2 + \varrho y)\, dx,$$

satisfying the boundary conditions

$$y(-l) = 0, \quad y'(-l) = 0, \quad y(l) = 0, \quad y'(l) = 0.$$

This variational problem arises in the study of the buckled axis of an elastic cylindrical beam with both ends fixed. If the beam is homogeneous, then ϱ and μ are constant and the Euler-Poisson equation is

$$\varrho + \frac{d^2}{dx^2}(\mu y'') = 0 \quad \text{or} \quad y^{\mathrm{iv}} = -\frac{\varrho}{\mu}$$

and hence

$$y = -\frac{\varrho x^4}{24\mu} + C_1 x^3 + C_2 x^2 + C_3 x + C_4.$$

From the boundary conditions we finally have

$$y = -\frac{\varrho}{24\mu}(x^4 - 2l^2 x^2 + l^4) \quad \text{or} \quad y = -\frac{\varrho}{24\mu}(x^2 - l^2)^2.$$

If the functional v has the form

$$v(y(x), z(x)) = \int_{x_0}^{x_1} F(x, y, y', \ldots, y^{(n)}, z, z', \ldots, z^{(m)})\, dx,$$

then varying only $y(x)$ and having kept $z(x)$ fixed, we find finally that any pair of functions $y(x)$, $z(x)$ that gives an extremum of this functional must satisfy the Euler-Poisson equation

$$F_y - \frac{d}{dx} F_{y'} + \ldots + (-1)^n \frac{d^n}{dx^n} F_{y^{(n)}} = 0,$$

and varying $z(x)$, having kept $y(x)$ fixed we obtain that both these functions should also satisfy the equation

$$F_z - \frac{d}{dx} F_{z'} + \ldots + (-1)^m \frac{d^m}{dx^m} F_{z^{(m)}} = 0.$$

Consequently, the functions $z(x)$ and $y(x)$ should satisfy the system of two equations

$$F_y - \frac{d}{dx} F_{y'} + \ldots + (-1)^n \frac{d^n}{dx^n} F_{y^{(n)}} = 0,$$

$$F_z - \frac{d}{dx} F_{z'} + \ldots + (-1)^m \frac{d^m}{dx^m} F_{z^{(m)}} = 0.$$

The same line of argument applies in the discussion of extrema of similar functionals depending on an arbitrary number of functions

$$v(y_1, y_2, \ldots, y_m) = \int_{x_0}^{x_1} F(x, y_1, y_1', \ldots, y_1^{(n_1)}, y_2, y_2', \ldots,$$

$$y_2^{(n_2)}, \ldots, y_m, y_m', \ldots, y_m^{(n_m)}) \, dx.$$

Varying any function $y_i(x)$, and keeping the remaining ones fixed, we find that the fundamental necessary condition for an extremum is

$$F_{y_i} - \frac{d}{dx} F_{y_i'} + \ldots + (-1)^{n_i} \frac{d^{n_i}}{dx^{n_i}} F_{y_i^{(n_i)}} = 0$$

$$(i = 1, 2, \ldots, m).$$

5. Functionals depending on functions of several independent variables

We now examine the extrema of the functional

$$v(z(x, y)) = \iint_D F\left(x, y, z, \frac{\partial z}{\partial x}, \frac{\partial z}{\partial y}\right) dx\, dy,$$

where the values of functions $z(x, y)$ on the boudary C of the domain D are prescribed, i. e. there is given a curve C^* in the space, and every admissible surface is supposed to pass through this curve (Fig. 16). We shall write $\partial z/\partial x = p$ and $\partial z/\partial y = q$. The function F is supposed to be differentiable, and the surface $z = z(x, y)$ giving an extremum is supposed to be differentiable twice.

We consider a one-parameter family of surfaces $z = z(x, y, a) = z(x, y) + a\,\delta z$, where $\delta z = z^*(x, y) - z(x, y)$. For $a = 0$, we have the surface $z = z(x, y)$ which has been just supposed to give an extremum, and for $a = 1$, a certain admissible surface $z = z^*(x, y)$, The functional considered only on the functions of the

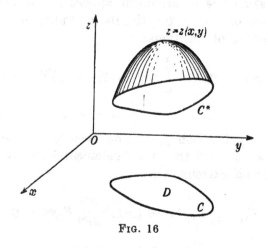

FIG. 16

family $z(x, y, a)$ turns into an ordinary function of one variable a, that should have an extremum at $a = 0$. Consequently

$$\frac{\partial}{\partial a}\, v\big(z(x, y, a)\big)\big|_{a=0} = 0.$$

Having called the derivative of $v\big(z(x, y, a)\big)$ with respect to a taken at $a = 0$ a variation of the functional, and having designated it by δv, as in Section 1, we find that

$$\delta v = \left[\frac{\partial}{\partial a}\iint_D F\big(x, y, z(x, y, a)\, p(x, y, a), q(x, y, a)\big)\,dx\,dy\right]_{a=0}$$
$$= \iint_D (F_z\,\delta z + F_p\,\delta p + F_q\,\delta q)\,dx\,dy,$$

where

$$z(x, y, a) = z(x, y) + a\,\delta z,$$

$$p(x, y, a) = \frac{\partial z(x, y, a)}{\partial x} = p(x, y) + a\,\delta p,$$

$$q(x, y, a) = \frac{\partial z(x, y, a)}{\partial y} = q(x, y) + a\,\delta q.$$

Since

$$\frac{\partial}{\partial x}\{F_p\,\delta z\} = \frac{\partial}{\partial x}\{F_p\}\,\delta z + F_p\,\delta p,$$

$$\frac{\partial}{\partial y}\{F_q\,\delta z\} = \frac{\partial}{\partial y}\{F_q\}\,\delta z + F_q\,\delta q,$$

it follows that

$$\iint_D (F_p\,\delta p + F_q\,\delta q)\,dxdy = \iint_D \left[\frac{\partial}{\partial x}\{F_p\,\delta z\} + \frac{\partial}{\partial y}\{F_q\,\delta z\}\right]dxdy$$

$$-\iint_D \left[\frac{\partial}{\partial x}\{F_p\} + \frac{\partial}{\partial y}\{F_q\}\right]\delta z\,dxdy,$$

where $\dfrac{\partial}{\partial x}\{F_p\}$ is a so called *total* partial derivative with respect to x. In computing it, y is considered constant whereas z, p, and q are still considered functions of x,

$$\frac{\partial}{\partial x}\{F_p\} = F_{px} + F_{pz}\frac{\partial z}{\partial x} + F_{pp}\frac{\partial p}{\partial x} + F_{pq}\frac{\partial q}{\partial x},$$

and similarly

$$\frac{\partial}{\partial y}\{F_q\} = F_{qy} + F_{qz}\frac{\partial z}{\partial y} + F_{qp}\frac{\partial p}{\partial y} + F_{qq}\frac{\partial q}{\partial y}.$$

Using Green's formula

$$\iint_D \left(\frac{\partial N}{\partial x} + \frac{\partial M}{\partial y}\right)dxdy = \int_C (N\,dy - M\,dx)$$

we have

$$\iint_D \left[\frac{\partial}{\partial x}\{F_p\,\delta z\} + \frac{\partial}{\partial y}\{F_q\,\delta z\} \right] dx\,dy = \int_C (F_p\,dy - F_q\,dx)\,\delta z = 0.$$

This latter integral vanishes, since $\delta z = 0$ along C, as all the admissible surfaces must pass through one and the same closed curve C in the space. Consequently,

$$\iint_D [F_p\,\delta p + F_q\,\delta q]\,dx\,dy = -\iint_D \left[\frac{\partial}{\partial x}\{F_p\} + \frac{\partial}{\partial y}\{F_q\} \right] \delta z\,dx\,dy$$

and the necessary conditions for an extremum

$$\iint_D (F_z\,\delta z + F_p\,\delta p + F_q\,\delta q)\,dx\,dy = 0$$

assumes the form

$$\iint_D \left(F_z - \frac{\partial}{\partial x}\{F_p\} - \frac{\partial}{\partial y}\{F_q\} \right) \delta z\,dx\,dy = 0.$$

Since the variation δz is arbitrary (δz has been subject to some restrictions of general kind such as continuity, differentiability, vanishing on the boundary curve C, and the like), and since the first factor is continuous, it follows from the fundamental lemma (p. 27) that if the surface $z = z(x, y)$ gives an extremum, then

$$F_z - \frac{\partial}{\partial x}\{F_p\} - \frac{\partial}{\partial y}\{F_q\} \equiv 0.$$

Consequently, $z(x, y)$ is a solution of the equation

$$F_z - \frac{\partial}{\partial x}\{F_p\} - \frac{\partial}{\partial y}\{F_q\} = 0.$$

This second-order partial differential equation, which is to be satisfied by each function giving an extremum is called the *Ostrogradski equation*, after the famous Russian mathematician M. B. Ostrogradski, who discovered it first in 1834. It is also sometimes known as the *Euler-Lagrange equation.*

EXAMPLE 1.

$$v\left(z\left(x,y\right)\right) = \int\int_D \left[\left(\frac{\partial z}{\partial x}\right)^2 + \left(\frac{\partial z}{\partial y}\right)^2\right] dxdy\,,$$

with a boundary condition that $z = f(x, y)$ on the boundary C of the domain D, where $f(x, y)$ is a given in advance function on the boundary curve C.

The Ostrogradski equation is

$$\frac{\partial^2 z}{\partial x^2} + \frac{\partial^2 z}{\partial y^2} = 0,$$

or briefly

$$\Delta z = 0,$$

i.e. it is the well-known Laplace equation. We have to find a solution of this equation that is continuous in D, and takes on the given in advance values on the boundary of D. This is one of the fundamental problems of mathematical physics; it is called the *Dirichlet problem.*

EXAMPLE 2.

$$v\left(z\left(x,y\right)\right) = \int\int_D \left[\left(\frac{\partial z}{\partial x}\right)^2 + \left(\frac{\partial z}{\partial y}\right)^2 + 2zf(x, y)\right] dxdy$$

where on the boundary of D the values of all the functions z are given in advance and fixed.

The Ostrogradski equation is

$$\frac{\partial^2 z}{\partial x^2} + \frac{\partial^2 z}{\partial y^2} = f(x, y),$$

or briefly

$$\Delta z = f(x, y).$$

This equation is called the *Poisson equation,* and it is also frequently encountered in various problems of mathematical physics.

EXAMPLE 3. *The problem of finding the minimum surface that passes through a given closed curve C in the space reduces to examining the minima of the functional*

$$S\left(z\left(x,y\right)\right) = \int\int_D \sqrt{1 + \left(\frac{\partial z}{\partial x}\right)^2 + \left(\frac{\partial z}{\partial y}\right)^2}\ dxdy.$$

The corresponding Ostrogradski equation is

$$\frac{\partial}{\partial x}\left\{\frac{p}{\sqrt{1+p^2+q^2}}\right\} + \frac{\partial}{\partial y}\left\{\frac{q}{\sqrt{1+p^2+q^2}}\right\} = 0$$

or

$$\frac{\partial^2 z}{\partial x^2}\left(1+\left(\frac{\partial z}{\partial y}\right)^2\right) - 2\frac{\partial z}{\partial x}\cdot\frac{\partial z}{\partial y}\cdot\frac{\partial^2 z}{\partial x\partial y} + \frac{\partial^2 z}{\partial y^2}\left(1+\left(\frac{\partial z}{\partial x}\right)^2\right) = 0.$$

For the functional

$$v\big(z(x_1, x_2, \ldots, x_n)\big)$$

$$= \int\int\ldots\int_D F(x_1, x_2, \ldots, x_n, z, p_1, p_2, \ldots, p_n)\, dx_1 dx_2 \ldots dx_n,$$

where $p_i = \partial z/\partial x_i$, we obtain from the fundamental necessary condition of an extremum, $\delta v = 0$, that the Ostrogradski equation has the form

$$F_z - \sum_{i=1}^{n}\frac{\partial}{\partial x_i}\{F_{p_i}\} = 0.$$

Any function

$$z = z(x_1, x_2, \ldots, x_n),$$

that makes v an extremum must satisfy this equation.

For instance, in the case of the functional

$$v = \int\int\int_D \left(\left(\frac{\partial u}{\partial x}\right)^2 + \left(\frac{\partial u}{\partial y}\right)^2 + \left(\frac{\partial u}{\partial z}\right)^2\right) dx dy dz$$

the Ostrogradski equation has the form

$$\frac{\partial^2 u}{\partial x^2} + \frac{\partial^2 u}{\partial y^2} + \frac{\partial^2 u}{\partial z^2} = 0.$$

If the integrand of the functional v depends on derivatives of higher order, then using successively the same line of argument as in derivation of the Ostrogradski equation we find that any function that makes v an extremum must satisfy an equation analogous to that of Euler-Poisson (p. 44-45).

For instance, in the case of

$$v\big(z(x, y)\big) = \iint_D F\left(x, y, z, \frac{\partial z}{\partial x}, \frac{\partial z}{\partial y}, \frac{\partial^2 z}{\partial x^2}, \frac{\partial^2 z}{\partial x \partial y}, \frac{\partial^2 z}{\partial y^2}\right) dx dy$$

we have the following equation

$$F_z - \frac{\partial}{\partial x}\{F_p\} - \frac{\partial}{\partial y}\{F_q\} + \frac{\partial^2}{\partial x^2}\{F_r\} + \frac{\partial^2}{\partial x \partial y}\{F_s\} +$$

$$+ \frac{\partial^2}{\partial y^2}\{F_t\} = 0,$$

where

$$p = \frac{\partial z}{\partial x}, \quad q = \frac{\partial z}{\partial y}, \quad r = \frac{\partial^2 z}{\partial x^2}, \quad s = \frac{\partial^2 z}{\partial x \partial y},$$

$$t = \frac{\partial^2 z}{\partial y^2}.$$

Any function that makes v an extremum must satisfy this fourth-order partial differential equation.

For instance, in case of

$$v = \iint_D \left(\left(\frac{\partial^2 z}{\partial x^2}\right)^2 + \left(\frac{\partial^2 z}{\partial y^2}\right)^2 + 2\left(\frac{\partial^2 z}{\partial x \partial y}\right)^2\right) dx dy,$$

any function that gives an extremum must satisfy the so-called *biharmonic equation*

$$\frac{\partial^4 z}{\partial x^4} + 2\frac{\partial^4 z}{\partial x^2 \partial y^2} + \frac{\partial^4 z}{\partial y^4} = 0,$$

which is usually written as $\Delta \Delta z = 0$.

For the functional

$$v = \iint_D \left(\left(\frac{\partial^2 z}{\partial x^2} \right)^2 + \left(\frac{\partial^2 z}{\partial y^2} \right)^2 + 2 \left(\frac{\partial^2 z}{\partial x \partial y} \right)^2 - 2zf(x, y) \right) dx dy$$

any function $z(x, y)$ that gives an extremum should satisfy the equation $\Delta \Delta z = f(x, y)$.

Likewise, the problem of finding extrema of the functional

$$v = \iint_D \left(\frac{\partial^2 z}{\partial x^2} + \frac{\partial^2 z}{\partial y^2} \right)^2 dx dy$$

or of a more general functional

$$v = \iint_D \left\{ \left(\frac{\partial^2 z}{\partial x^2} + \frac{\partial^2 z}{\partial y^2} \right)^2 - \right.$$

$$\left. - 2(1 - \mu) \left[\frac{\partial^2 z}{\partial x^2} \cdot \frac{\partial^2 z}{\partial y^2} - \left(\frac{\partial^2 z}{\partial x \partial y} \right)^2 \right] \right\} dx dy,$$

where μ is a parameter, lead to the investigation of the biharmonic equation.

6. Parametric representation of variational problems

In many variational problems it is most convenient to obtain a solution in a parametric representation. For instance, this is the case for the isoperimetric problem (see p. 12), in which we find a closed curve of a given perimeter l and maximal area S. It is not convenient to try to find a solution of the form $y = y(x)$, for it follows then from the very nature of the problem that $y(x)$ is not a univalent function (Fig. 17). It is therefore reasonable to try to find the solution in a parametric form $x = x(t)$, $y = y(t)$. Consequently, in this case, we must find extrema of the functional

$$S(x(t), y(t)) = \tfrac{1}{2} \int_0^T (xy^{\cdot} - yx^{\cdot}) dt$$

with the additional condition that $l = \int_0^T \sqrt{x^{\cdot 2} + y^{\cdot 2}} \, dt$, where l is a constant.

Suppose that in examining the extrema of a functional

$$v(y(x)) = \int\limits_{x_0}^{x_1} F(x, y, y') \, dx$$

it has turned out that it is better to try to find the solution in a parametric form, $x = x(t), y = y(t)$. Then the functional turns into

$$v(x(t), y(t)) = \int\limits_{t_0}^{t_1} F\left(x(t), y(t), \frac{y^{\cdot}(t)}{x^{\cdot}(t)}\right) x^{\cdot}(t) \, dt.$$

We observe that the transformed integrand function

$$F\left(x(t), y(t), \frac{y^{\cdot}(t)}{x^{\cdot}(t)}\right) x^{\cdot}(t)$$

does not depend on t explicitly and it is a homogeneous function of the first order with respect to the variables $x^{\cdot}$ and $y^{\cdot}$.

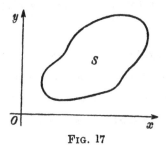

Fig. 17

Consequently, the functional $v(x(t), y(t))$ is not an arbitrary one of the form

$$\int\limits_{t_0}^{t_1} \varPhi(t, x(t), y(t), x^{\cdot}(t), y^{\cdot}(t)) \, dt,$$

depending on two functions $x(t)$ and $y(t)$. Conversely, it is only a rather special particular case of such functional, for the integrand does not depend explicitly on t and furthermore, it is a first-order homogeneous function with respect to $x^{\cdot}$ and $y^{\cdot}$.

If we introduce some other parametric representation $x = x(\tau)$ $y = y(\tau)$, then the functional $v(x, y)$ will turn into

$$\int\limits_{\tau_0}^{\tau_1} F\left(x, y, \frac{y_{\tau}^{\cdot}}{x_{\tau}^{\cdot}}\right) x_{\tau}^{\cdot} \, d\tau.$$

Consequently, this special form of the integrand function of v does not depend on the parametric representation chosen. The functional v depends only on the curve along which the integration is performed, and it does not depend on the parametric representation of this curve.

It is not difficult to show that if the integrand of the functional

$$v(x(t),\, y(t)) = \int_{t_0}^{t_1} \Phi(t,\, x(t),\, y(t),\, x^{\cdot}(t),\, y^{\cdot}(t))\, dt$$

does not depend on t explicitly and if it is a first-order homogeneous function with respect to $x^{\cdot}$ and $y^{\cdot}$, then the functional $v(x(t),\, y(t))$ depends only on the curve $x = x(t)$, $y = y(t)$ and it does not depend on any particular choice of the parametric representation of this curve.

In fact, suppose

$$v(x(t),\, y(t)) = \int_{t_0}^{t_1} \Phi(x(t),\, y(t),\, x^{\cdot}(t),\, y^{\cdot}(t))\, dt,$$

where

$$\Phi(x,\, y,\, kx^{\cdot},\, ky^{\cdot}) = k\, \Phi(x,\, y,\, x^{\cdot},\, y^{\cdot}).$$

We will pass to some other parametric representation, setting

$$\tau = \varphi(t), \quad \text{where} \quad \varphi^{\cdot}(t) \neq 0,$$

$$x = x(\tau), \quad y = y(\tau).$$

Then

$$\int_{t_0}^{t_1} \Phi(x(t),\, y(t),\, x^{\cdot}(t),\, y^{\cdot}(t))\, dt$$

$$= \int_{\tau_0}^{\tau_1} \Phi\left(x(\tau),\, y(\tau),\, x_{\tau}^{\cdot}(t),\, y_{\tau}^{\cdot}\varphi^{\cdot}(t)\right) \frac{d\tau}{\varphi^{\cdot}(t)}.$$

Since Φ is a first-order homogeneous function with respect to $x^{\cdot}$ and $y^{\cdot}$ we have

$$\Phi(x,\, y,\, x_{\tau}^{\cdot}\varphi^{\cdot},\, y_{\tau}^{\cdot}\varphi^{\cdot}) = \varphi^{\cdot}\, \Phi(x,\, y,\, x_{\tau}^{\cdot},\, y_{\tau}^{\cdot}),$$

and consequently

$$\int_{t_0}^{t_1} \Phi(x,\, y,\, x_{t}^{\cdot},\, y_{t}^{\cdot})\, dt = \int_{\tau_0}^{\tau_1} \Phi(x,\, y,\, x_{\tau}^{\cdot},\, y_{\tau}^{\cdot})\, d\tau,$$

i.e. the integrand remains unchanged under the transformation of the parameter.

Examples of such functionals are the length of a curve $\int_{t_0}^{t_1} \sqrt{x^{\cdot 2} + y^{\cdot 2}}\, dt$ (1), and the area encircled by a curve $\frac{1}{2}\int_{t_0}^{t_1} (xy^{\cdot} - yx^{\cdot})\, dt$.

In order to find the extremals of functionals of the type

$$v(x(t),\, y(t)) = \int_{t_0}^{t_1} \Phi(x,\, y,\, x^{\cdot},\, y^{\cdot})\, dt,$$

where Φ is a first-order homogeneous function with respect to $x^{\cdot}$ and $y^{\cdot}$, we have to solve the system of Euler equations

$$\Phi_x - \frac{d}{dt}\, \Phi_{x^{\cdot}} = 0, \quad \Phi_y - \frac{d}{dt}\, \Phi_{y^{\cdot}} = 0,$$

as has to be done for function 's with arbitrary integrand $\Phi(t,\, x,\, y,\, x^{\cdot},\, y^{\cdot})$.

However, in this particular case these equations are no longer independent, for along with a certain solution $x = x(t)$, $y = y(t)$ there are some other pairs of functions, giving other parametric representation of this same curve, and all these pairs must satisfy the Euler equations too, and, if these equations were independent, this would be a contradiction, because of a general theorem of existence and uniqueness of solution of a system of differential equations. Consequently, for a functional of the form

$$v(x(t),\, y(t)) = \int_{t_0}^{t_1} \Phi(x,\, y,\, x^{\cdot},\, y^{\cdot})\, dt$$

where Φ is a first-order homogeneous function with respect to $x^{\cdot}$ and $y^{\cdot}$, one of the Euler equations is a consequence of the other. To find extremals, it is therefore sufficient to solve one of them along with an equation determining the parameter. For instance, the equation $\Phi_x - \dfrac{d}{dt}\, \Phi_{x^{\cdot}} = 0$ can be solved along with $x^{\cdot 2} + y^{\cdot 2} = 1$, this latter equation expressing the fact that the parameter chosen was the arc-length of a curve.

---

(1) The function $\sqrt{x^{\cdot 2} + y^{\cdot 2}}$ is a positively homogeneous function of the first order, i.e. it satisfies the relation $F(kx,\, ky) = k^{\prime} F(x,\, y)$ for positive $k^{\prime}$ only. Nevertheless, this is quite sufficient for the theory outlined in this section, for when transforming the parameter, $\tau = \varphi(t)$, it can be assumed with no loss of generality that $\varphi^{\cdot}(t) > 0$.

7. Some applications

The fundamental variational principle of mechanics is the *principle of Ostrogradski-Hamilton*, that among all possible motions of a system of material points, i.e. among those compatible with the constraints, the actual motion traces out the curve that gives an extremum of the integral

$$\int_{t_0}^{t_1} (T - U)\, dt,$$

where T is the kinetic and U, potential energy of the system.

Let us apply this principle to some mechanical problems.

EXAMPLE 1. *Given a system of material points with masses m_i, $i = 1, 2, \ldots, n$, and coordinates (x_i, y_i, z_i), on each point acts a corresponding force F_i that is determined by a force function U, a potential, depending on the coordinates only:*

$$F_{ix} = -\frac{\partial U}{\partial x_i}, \qquad F_{iy} = -\frac{\partial U}{\partial y_i}, \qquad F_{iz} = -\frac{\partial U}{\partial z_i},$$

where F_{ix}, F_{iy}, F_{iz} are the coordinates of vector F_i acting on the point (x_i, y_i, z_i). Find the differential equations of motion of the system.

The kinetic energy is

$$T = \tfrac{1}{2} \sum_{i=1}^{n} m_i (\dot{x}_i^2 + \dot{y}_i^2 + \dot{z}_i^2),$$

and the potential energy of the system is U. The Euler equations for $\int_{t_0}^{t_1} (T - U)\, dt$ are

$$-\frac{\partial U}{\partial x_i} - \frac{d}{dt} \cdot \frac{\partial T}{\partial \dot{x}_i} = 0,$$

$$-\frac{\partial U}{\partial y_i} - \frac{d}{dt} \cdot \frac{\partial T}{\partial \dot{y}_i} = 0,$$

$$-\frac{\partial U}{\partial z_i} - \frac{d}{dt} \cdot \frac{\partial T}{\partial \dot{z}_i} = 0,$$

or

$$m_i x_i^{..} - F_{ix} = 0,$$

$$m_i y_i^{..} - F_{iy} = 0,$$

$$m_i z_i^{..} - F_{iz} = 0, \quad i = 1, 2, \ldots, n.$$

If the system is subject to independent constraints of the form

$$\varphi_j(t, x_1, x_2, \ldots, x_n, y_1, y_2, \ldots, y_n, z_1, z_2, \ldots, z_n) = 0,$$

$$j = 1, 2, \ldots, m, \quad m < 3n,$$

then from these equations, m variables can be expressed by the $3n-m$ independent variables exclusive of the time t, or all $3n$ variables can be replaced by $3n-m$ new variables which are independent

$$q_1, q_2, \ldots, q_{3n-m}.$$

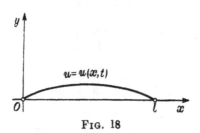

FIG. 18

Now T and U can be considered as functions of $q_1, q_2, \ldots, q_{3n-m}$, and t,

$$T = T(q_1, q_2, \ldots, q_{3n-m}, q_1^{.}, q_2^{.}, \ldots, q_{3n-m}^{.}, t),$$

$$U = U(q_1, q_2, \ldots, q_{3n-m}, t),$$

and the Euler equations are

$$\frac{\partial(T-U)}{\partial q_i} - \frac{d}{dt} \frac{\partial T}{\partial q_i^{.}} = 0, \quad i = 1, 2, \ldots, 3n-m.$$

(*Lagrange's equation of motion*).

EXAMPLE 2. *Derive a differential equation describing the free vibrations of a string.*

We place the origin of the coordinate system at one of the end points of the string. The string in equilibrium is stretched along a straight line. We choose this line as the x-axis (Fig. 18). The deflection $u(x, t)$ of the point x from this equilibrium position is a function of x and t.

The potential energy U of an element of a perfectly elastic string is proportional to a longitudinal displacement. The piece of string dx in the deformed state has the length $ds = \sqrt{1 + u_x^2}\, dx$, up to infinitesimal of higher order, and consequently it is $(\sqrt{1 + u_x^2} - 1)\, dx$ longer. By Taylor's formula, $\sqrt{1 + u_x^2} \approx 1 + \tfrac{1}{2} u_x^2$.

Considering u_x to be small and neglecting the higher powers of u_x, the potential energy of the piece dx of the string is $\tfrac{1}{2} k u_x^2\, dx$, where k is a certain coefficient, and the potential energy of the whole string is

$$\tfrac{1}{2} \int\limits_0^l k u_x^2\, dx.$$

The integral $\int\limits_{t_0}^{t_1} (T - U)\, dt$ has the form

$$v = \int\limits_{t_0}^{t_1} \int\limits_0^l \left(\tfrac{1}{2} \varrho u_t^2 - \tfrac{1}{2} k u_x^2 \right) dx\, dt.$$

The equation of motion of the string is an Ostrogradski equation for the functional v. Consequently, the equation of motion of a string is

$$\frac{\partial}{\partial t} \left(\varrho\, \frac{\partial u}{\partial t} \right) - \frac{\partial}{\partial x} \left(k\, \frac{\partial u}{\partial x} \right) = 0.$$

If the string is homogeneous, then ϱ and k are constants, and the equation of such vibrating string reduces to

$$\varrho\, \frac{\partial^2 u}{\partial t^2} - k\, \frac{\partial^2 u}{\partial x^2} = 0.$$

Suppose now that there is an external force $f(t, x)$ acting on the string. This force is assumed to be perpendicular to the string in its equilibrium position and to be proportional to the mass of an element of string. It is easy to see that such an external force acting on an element of the string does work which can be expressed in the form $\varrho f(t, x) u\, dx$. Consequently, the Ostrogradski-Hamilton integral $\int\limits_{t_0}^{t_1} (T - U)\, dt$ has the form

$$\int\limits_{t_0}^{t_1} \int\limits_0^l \left(\tfrac{1}{2} \varrho u_t^2 - \tfrac{1}{2} k u_x^2 + \varrho f(t, x) u \right) dx\, dt,$$

and hence the equation of forced vibrations of a string is

$$\frac{\partial}{\partial t} \left(\varrho u_t \right) - \frac{\partial}{\partial x} \left(k u_x \right) - \varrho f(t, x) = 0,$$

or in the case of a string which is homogeneous

$$\frac{\partial^2 u}{\partial t^2} - \frac{k}{\varrho} \cdot \frac{\partial^2 u}{\partial x^2} = f(t, x).$$

The equation of a vibrating membrane can be derived in a similar way.

EXAMPLE 3. *Derive the equation of a vibrating linear rod.*

We choose the longitudinal axis in the position of equilibrium as the x-axis. The deflection from the rest position $u(x, t)$ being a function of x and the time t, the kinetic energy of the rod is

$$T = \tfrac{1}{2} \int_0^l \varrho u_t^2 \, dx,$$

where l is its length. We assume that the rod will not stretch. The potential energy of an elastic rod is then proportional to the square of its curvature. Consequently,

$$U = \tfrac{1}{2} \int_0^l k \left\{ \frac{\partial^2 u / \partial x^2}{[1 + (\partial u / \partial x)^2]^{3/2}} \right\}^2 dx.$$

Suppose the deviation from the resting position small, so that the term $(\partial u / \partial x)^2$ in denominator can be dropped. Consequently,

$$U = \tfrac{1}{2} \int_0^l k \left(\frac{\partial^2 u}{\partial x^2} \right)^2 dx.$$

The integral of Ostrogradski-Hamilton then has the form

$$\int_{t_0}^{t_1} \int_0^l \left[\tfrac{1}{2} \varrho u_t^2 - \tfrac{1}{2} k u_{xx}^2 \right] dx dt.$$

Therefore the equation of motion of free vibrations of an elastic rod is

$$\frac{\partial}{\partial t} (\varrho u_t) + \frac{\partial^2}{\partial x^2} (k u_{xx}) = 0.$$

If the rod is homogeneous, then ϱ and k are constants, and the equation of homogeneous vibrating rod turns into

$$\varrho \frac{\partial^2 u}{\partial t^2} + k \frac{\partial^4 u}{\partial x^4} = 0.$$

If there is an external force $f(x, t)$, acting on the rod, then one shall make allowance for the potential of this force (cf. the previous example).

Problems

Find the extremals of the functionals:

1. $v(y(x)) = \int\limits_{x_0}^{x_1} \dfrac{\sqrt{1+y'^2}}{y}\, dx$;

2. $v(y(x)) = \int\limits_{x_0}^{x_1} (y^2 + 2xyy')\, dx$, $\quad y(x_0) = y_0$, $\quad y(x_1) = y_1$;

3. $v(y(x)) = \int\limits_{0}^{1} (xy + y^2 - 2y^2 y')\, dx$, $\quad y(0) = 1$. $\quad y(1) = 2$;

4. $v(y(x)) = \int\limits_{x_0}^{x_1} \sqrt{y(1+y'^2)}\, dx$;

5. $v(y(x)) = \int\limits_{x_0}^{x_1} y'(1 + x^2 y')\, dx$;

6. $v(y(x)) = \int\limits_{x_0}^{x_1} (y'^2 + 2yy' - 16y^2)\, dx$;

7. $v(y(x)) = \int\limits_{x_0}^{x_1} (xy' + y'^2)\, dx$;

8. $v(y(x)) = \int\limits_{x_0}^{x_1} \dfrac{1+y^2}{y'^2}\, dx$;

9. $v(y(x)) = \int\limits_{x_0}^{x_1} (y^2 + y'^2 - 2y\sin x)\, dx$;

10. $v(y(x)) = \int\limits_{x_0}^{x_1} (16y^2 - y''^2 + x^2)\, dx$;

11. $v(y(x)) = \int\limits_{x_0}^{x_1} (2xy + y'''^2)\, dx$;

12. $v(y(x), z(x)) = \int\limits_{x_0}^{x_1} (2yz - 2y^2 + y'^2 - z'^2)\, dx$;

13. $v(y(x), z(x)) = \int\limits_{x_0}^{x_1} (y'^2 + z'^2 + y'z')\, dx$;

14. Write down the Ostrogradski equation for the functional

$$v(z(x, y)) = \iint\limits_{D} \left(\left(\frac{\partial z}{\partial x} \right)^2 - \left(\frac{\partial z}{\partial y} \right)^2 \right) dx\, dy.$$

15. Write down the Ostrogradski equation for the functional

$$v\big(u(x,y,z)\big) = \iiint\limits_{D} \left(\left(\frac{\partial u}{\partial x}\right)^2 + \left(\frac{\partial u}{\partial y}\right)^2 + \left(\frac{\partial u}{\partial z}\right)^2 + \right.$$

$$\left. + 2uf(x,y,z)\right) dx\,dy\,dz.$$

Find the extremals of the functionals:

16. $v\big(y(x)\big) = \displaystyle\int_{x_0}^{x_1} \frac{y'^2}{x^3}\, dx.$

17. $v\big(y(x)\big) = \displaystyle\int_{x_0}^{x_1} (y^2 + 2y'^2 + y''^2)\, dx.$

18. $v\big(y(x)\big) = \displaystyle\int_{x_0}^{x_1} (y^2 - y'^2 - 2y\cosh x)\, dx.$

19. $v\big(y(x)\big) = \displaystyle\int_{x_0}^{x_1} (y^2 + y'^2 + 2ye^x)\, dx.$

20. $v\big(y(x)\big) = \displaystyle\int_{x_0}^{x_1} (y^2 - y'^2 - 2y\sin x)\, dx.$

CHAPTER II

VARIATIONAL PROBLEMS WITH MOVABLE BOUNDARIES AND SOME OTHER PROBLEMS

1. Simplest problem with movable boundaries

It was assumed, in Chapter I, that the end points (x_0, y_0) and (x_1, y_1) of the functional

$$v = \int_{x_0}^{x_1} F(x, y, y') \, dx$$

are fixed in advance. We shall now suppose that one or both of the end points are variable, and consequently the class of admissible curves will expand—along with such comparison curves that have the same end points as the curve under examination, we shall also consider curves with arbitrary end points.

Therefore if a curve $y = y(x)$ gives an extremum for a problem with variable boundary points, then of course, the same curve gives an extremum with respect to a more restricted class of curves having the same boundary points as the curve $y = y(x)$, and consequently the fundamental necessary condition of an extremum for a problem with fixed boundary point should hold. The function $y(x)$ should be a solution of Euler equation

$$F_y - \frac{d}{dx} F_{y'} = 0.$$

Consequently, the curves $y = y(x)$ that give extrema for a problem with moving boundaries should be extremals.

The general solution of Euler equation involves two

arbitrary constants. To determine them, two additional relations are required. In the case of a problem with fixed boundaries these relations were

$$y(x_0) = y_0 \quad \text{and} \quad y(x_1) = y_1.$$

In the case of movable boundaries one or both of these relations do not hold, so to determine the arbitrary constants in the general solution of Euler equation, some other conditions must be obtained from the fundamental necessary condition for an extremum (the condition that δv vanishes).

Since for a problem with movable boundaries an extremum can occur only when the curve is one of the integral curves $y = y(x, C_1, C_2)$ of the Euler equation, we shall direct our attention only to those values of the functional that are taken along these integral curves. We can therefore consider the functional v to be a function $v\big(y(x, C_1, C_2)\big)$ of two parameters C_1 and C_2 and of the limits of integration x_0, x_1, so that the variation of that functional coincides with the differential of this function. For the sake of simplicity, we assume that one of the end points, for instance (x_0, y_0), is fixed while the other (x_1, y_1) can move turning into $(x_1 + \Delta x_1, y_1 + \Delta y_1)$, or as it is usually designated in the calculus of variations, $(x_1 + \delta x_1, y_1 + \delta y_1)$.

Two admissible curves $y = y(x)$ and $y = y(x) + \delta y$ will be considered close to each other, if the absolute values of variations δy and $\delta y'$ are small and so are the absolute values of increments δx_1 and δy_1 (the increments δx_1 and δy_1 are usually called *variations of boundary values* x_1 and y_1, respectively).

The extremals passing through the point (x_0, y_0) form a pencil of extremals $y = y(x, C_1)$. The functional $v\big(y(x, C_1)\big)$, restricted to the curves belonging to this pencil, becomes a function of C_1 and x_1. If the curves that belong to this pencil do not intersect in a neighbourhood of an extremal under consideration, then $v\big(y(x, C_1)\big)$

can be thought of as a univalent function with respect to x_1 and y_1, for an arbitrary choice of x_1 and y_1 determines uniquely an extremal of the pencil (Fig. 19) and by the same token a value of the functional.

Let us calculate the variation of the functional $v\big(y(x, C_1)\big)$ taken along extremals of the pencil $y = y(x, C_1)$, under the assumption that the end point (x_1, y_1) moves to $(x_1 + \delta x_1, y_1 + \delta y_1)$. Since the functional taken only along the curves of the pencil turns into a function of x_1 and y_1, its variation turns into the differential of this function.

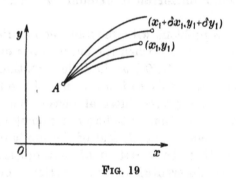

FIG. 19

We separate the main linear part of the increment Δv, that is linear with respect to δx_1 and δy_1,

$$
\text{(I)} \quad \Delta v = \int_{x_0}^{x_1 + \delta x_1} F(x, y + \delta y, y' + \delta y')\, dx - \int_{x_0}^{x_1} F(x, y, y')\, dx
$$

$$
= \int_{x_1}^{x_1 + \delta x_1} F(x, y + \delta y, y' + \delta y')\, dx +
$$

$$
+ \int_{x_0}^{x_1} \big(F(x, y + \delta y, y' + \delta y') - F(x, y, y')\big)\, dx.
$$

We shall transform the first term of the right-hand side with the aid of the mean value theorem

$$
\int_{x_1}^{x_1 + \delta x_1} F(x, y + \delta y, y' + \delta y')\, dx = F\big|_{x = x_1 + \theta \delta x_1}\, \delta x_1,
$$

where $0 < \theta < 1$. Furthermore, by virtue of continuity of the function F

$$F|_{x=x_1+\theta\delta x_1} = F(x, y, y')|_{x=x_1} + \varepsilon_1,$$

where $\varepsilon_1 \to 0$, as $\delta x_1 \to 0$ and $\delta y_1 \to 0$.

Consequently,

$$\int_{x_1}^{x_1+\delta x_1} F(x, y+\delta y, y'+\delta y')\,dx = F(x, y, y')|_{x=x_1}\delta x_1 + \varepsilon_1\, dx_1.$$

To transform the second term of the right-hand side of (I), we shall utilize the Taylor formula

$$\int_{x_0}^{x_1} \big(F(x, y+\delta y, y'+\delta y') - F(x, y, y')\big)\,dx$$

$$= \int_{x_0}^{x_1} \big(F_y(x, y, y')\,\delta y + F_{y'}(x, y, y')\,\delta y'\big)\,dx + R_1,$$

where R_1 is an infinitesimal of higher order than δy or $\delta y'$. Further, integrating by parts the second term of its integrand, the linear part of the increment Δv

$$\int_{x_0}^{x_1} (F_y\,\delta y + F_{y'}\,\delta y')\,dx$$

can be transformed to

$$[F_{y'}\,\delta y]_{x_0}^{x_1} + \int_{x_0}^{x_1} \left(F_y - \frac{d}{dx}\,F_{y'}\right)\delta y\,dx.$$

The values of the functional are taken only along extremals, and consequently

$$F_y - \frac{d}{dx}\,F_{y'} \equiv 0.$$

Since the end point (x_0, y_0) is fixed, it follows that $\delta y|_{x=x_0} = 0$ and therefore

$$\int_{x_0}^{x_1} (F_y\,\delta y + F_{y'}\,\delta y')\,dx = [F_{y'}\,\delta y]_{x=x_1}.$$

Observe that $\delta y|_{x=x_1}$ does not mean the same as δy_1, the increment of y_1, for δy_1 is the change of y-coordinate of the free end point when it is moved from (x_1, y_1) to $(x_1 + \delta x_1, y_1 + \delta y_1)$, whereas $\delta y|_{x=x_1}$ is the change of y-coordinate of an extremal produced at the point $x = x_1$ when this extremal changes from one that passes through the points (x_0, y_0) and (x_1, y_1) to another one passing through (x_0, y_0) and $(x_1 + \delta x_1, y_1 + \delta y_1)$ (Fig. 20).

As it is shown at Fig. 20,

$$BD = \delta y|_{x=x_1}, \quad FC = \delta y_1, \quad EC \approx y'(x_1)\,\delta x_1,$$

$$BD = FC - EC$$

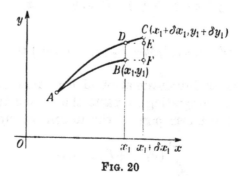

FIG. 20

or apart from infinitesimals of higher order

$$\delta y|_{x=x_1} \approx \delta y_1 - y'(x_1)\,\delta x_1.$$

Consequently we have

$$\int_{x_1}^{x_1 + \delta x_1} F\,dx \approx F|_{x=x_1}\,\delta x_1,$$

$$\int_{x_0}^{x_1} \big(F(x, y + \delta y, y' + \delta y') - F(x, y, y')\big)\,dx$$

$$\approx F_{y'}|_{x=x_1}\big(\delta y_1 - y'(x_1)\delta x_1\big),$$

where all approximate equations hold apart from infinitesimals of second or higher order with respect to δx_1

or δy_1. And so it follows from (I) that

$$\delta v = F\big|_{x=x_1} \delta x_1 + F_{y'}\big|_{x=x_1} \big(\delta y_1 - y'(x_1)\,\delta x_1\big)$$
$$= (F - y'F_{y'})\big|_{x=x_1} \delta x_1 + F_{y'}\big|_{x=x_1} \delta y_1 ,$$

or

$$dv^*(x_1, y_1) = (F - y'F_{y'})\big|_{x=x_1} dx_1 + F_{y'}\big|_{x=x_1} dy_1 ,$$

where $v^*(x, y)$ is a function to which the functional v reduces when taken only along extremals of the pencil $y = y(x, C_1)$ and $dx_1 = \Delta x_1 = \delta x_1$, $dy_1 = \Delta y_1 = \delta y_1$ are the changes in coordinates of the variable end point. The fundamental necessary condition for an extremum $\delta v = 0$ takes the form

(II) $\qquad (F - y'F_{y'})\big|_{x=x_1}\delta x_1 + F_{y'}\big|_{x=x_1}\delta y_1 = 0 .$

If the variations δx_1 and δy_1 are independent, then it follows that

$$(F - y'F_{y'})\big|_{x=x_1} = 0 \quad \text{and} \quad F_{y'}\big|_{x=x_1} = 0 .$$

However, it is often required that the variations δx_1 and δy_1 be dependent.

For instance, suppose that the right end point (x_1, y_1) can move along a certain curve

$$y_1 = \varphi(x_1).$$

It follows then that $\delta y_1 \approx \varphi'(x_1)\,\delta x_1$, and consequently condition (II) turns into $\big(F + (\varphi' - y')F_{y'}\big)\delta x_1 = 0$ or, since δx_1 varies arbitrarily, $\big(F + (\varphi' - y')F_{y'}\big)\big|_{x=x_1} = 0$. This condition establishes some relation between directional coefficients φ' and y' at the end point. It is called the *transversality condition*.

The transversality condition along with the relation $y_1 = \varphi(x_1)$ enables us, in general, to distinguish one or more extremals from the pencil $y = y(x, C)$, that can give an extremum. If the end point (x_0, y_0) can move along a curve $y = \psi(x)$, then by the same argument we obtain that

$$[F + (\psi' - y')F_{y'}]_{x=x_0} = 0,$$

which is the transversality condition at the point (x_0, y_0).

EXAMPLE 1. *Find the transversality condition for functionals of the form*

$$v = \int_{x_0}^{x_1} A(x, y)\sqrt{1+y'^2}\, dx.$$

The transversality condition $F + F_{y'}(\varphi' - y') = 0$ takes the form

$$A(x, y)\sqrt{1+y'^2} + \frac{A(x, y)y'}{\sqrt{1+y'^2}}\,(\varphi' - y') = 0$$

or

$$\frac{A(x, y)(1 + \varphi'y')}{\sqrt{1+y'^2}} = 0.$$

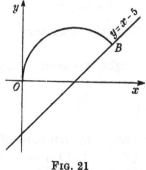 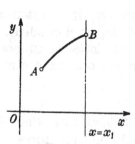

FIG. 21   FIG. 22

If we assume that $A(x, y) \neq 0$ at the end point, then we have $1 + y'\varphi' = 0$ or $y' = -1/\varphi'$, which means that in this particular case the transversality condition reduces to the orthogonality condition.

EXAMPLE 2. *Examine extrema of the functional*

$$\int_0^{x_1} \frac{\sqrt{1-y'^2}}{y}\, dx, \quad \text{where} \quad y(0) = 0, \quad \text{and} \quad y_1 = x_1 - 5$$

(Fig. 21). The integral curves of the Euler equation are the circles $(x-C_1)^2+y^2=C_2^2$ (Problem 1, p. 62). The first boundary condition yields $C_1=C_2$. Since for the given functional the transversality condition reduces to orthogonality condition (cf. the previous example), it follows that the straight line $y_1=x_1-5$ should be a diameter, and consequently the centre of the circle wanted is $(0,5)$, which is the intersection of the straight line $y_1=x_1-5$ with the x-axis. Hence $(x-5)^2+y^2=25$, or $y=\pm\sqrt{10x-x^2}$. Therefore an extremum can occur only along the path $y=\sqrt{10x-x^2}$ or $y=-\sqrt{10x-x^2}$.

If the end point (x_1, y_1) can move along a vertical line only (Fig. 22), so that $\delta x_1=0$, then relation (II) reduces to $F_{y'}|_{x=x_1}=0$. For instance, suppose that in a brachistochrone problem (cf. p. 38) the left end point is fixed, while the right one can move along a vertical line.

The extremals of the functional

$$v=\int_0^{x_1}\frac{\sqrt{1+y'^2}}{\sqrt{y}}\,dx$$

are cycloids. With the condition $y(0)=0$ in mind, the parametric equations of these cycloids are

$$x=C_1(t-\sin t),\qquad y=C_1(1-\cos t).$$

In order to determine C_1, we shall make use of the condition $F_{y'}|_{x=x_1}=0$ which in this case takes the form

$$\frac{y'}{\sqrt{y}\,\sqrt{1+y'^2}}\Bigg|_{x=x_1}=0,$$

whence $y'(x_1)=0$. The wanted cycloid should therefore intersect the vertical line $x=x_1$ at right angle, and consequently the point $x=x_1$, $y=y_1$ should be the lowest point of the cycloid (Fig. 23). Since the value $t=\pi$ of the parameter corresponds to this point, it follows that $x_1=C_1\pi$, $C_1=x_1/\pi$. Therefore an extremum can occur only along the cycloid

$$x = \frac{x_1}{\pi}(t - \sin t),\; y = \frac{x_1}{\pi}(1 - \cos t).$$

If the end point (x_1, y_1) occurring in a problem of the extrema of a functional $v = \int_{x_0}^{x_1} F(x, y, y')\,dx$ can move along a horizontal line $y = y_1$, then $\delta y_1 = 0$ and therefore

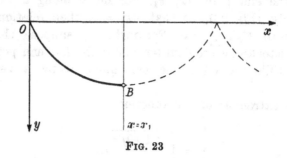

FIG. 23

relation (II) or the transversality condition takes the form

$$[F - y' F_{y'}]_{x=x_1} = 0.$$

2. Problems with movable boundaries for functionals of the form $\int_{x_0}^{x_1} F(x, y, z, y', z')\,dx$

If by investigating the extrema of a functional

$$v = \int_{x_0}^{x_1} F(x, y, z, y', z')\,dx$$

one of the boundary points, say $B(x_1, y_1, z_1)$, can move while the other $A(x_0, y_0, z_0)$ is fixed, or both boundary points can move, then it is evident that an extremum can be attained only along integral curves of the Euler equations

$$F_y - \frac{d}{dx} F_{y'} = 0, \quad F_z - \frac{d}{dx} F_{z'} = 0.$$

In fact, suppose that some curve C gives an extremum for a problem with movable boundaries. In other words, along C a maximum or a minimum value of v is attained with respect to all neighbouring admissible curves. Among these curves there are those which have their end points in common with the curve C giving an extremum. Consequently, it follows again that there is an extremum along C with respect to a more restricted class of neighbouring curves, namely those that have their end points in common with the curve C.

The curve C should therefore satisfy the necessary condition of an extremum for a problem with fixed boundary points. In particular, it should be an integral curve of the system of Euler equations.

The general solution of the system of Euler equations involves four arbitrary constants. If the coordinates of one of the boundary points $A(x_0, y_0, z_0)$ that is considered fixed are given in advance, then, in general, two of these arbitrary constants could be eliminated.

In order to evaluate the remaining two arbitrary constants, two further relations are required. These relations shall be obtained from the condition $\delta v = 0$. Calculating this variation we will now assume that the functional has been restricted to the integral curves of the system of Euler equations, for an extremum can be attained only along these curves. Thus, the functional v becomes a function $\Phi(x_1, y_1, z_1)$ of the coordinates x_1, y_1, z_1 of the point $B(x_1, y_1, z_1)$, and the variation of this functional turns into the differential of this function [1].

---

[1] This function Φ is univalent, provided the extremals from the pencil passing through the point A do not intersect at any neighbouring point, for in such case the point $B(x_1, y_1, z_1)$ always determines a unique extremal that passes through it.

By a similar calculation to that carried out above (p. 66-69)

$$\Delta v = \int\limits_{x_0}^{x_1+\delta x_1} F(x,\, y+\delta y,\, z+\delta z,\, y'+\delta y',\, z'+\delta z')\, dx -$$

$$- \int\limits_{x_0}^{x_1} F(x,\, y,\, z,\, y',\, z')\, dx$$

$$= \int\limits_{x_1}^{x_1+\delta x_1} F(x,\, y+\delta y,\, z+\delta z,\, y'+\delta y',\, z'+\delta z')\, dx +$$

$$+ \int\limits_{x_0}^{x_1} \big(F(x,\, y+\delta y,\, z+\delta z,\, y'+\delta y',\, z'+\delta z') -$$

$$- F(x,\, y,\, z,\, y',\, z')\big)\, dx.$$

We apply the mean value theorem to the first of these integrals and refer to its continuity, and by the Taylor formula we separate the main linear part from the second integral. We then have

$$\delta v = F|_{x=x_1} \delta x_1 + \int\limits_{x_0}^{x_1} (F_y\, \delta y + F_z\, \delta z + F_{y'}\, \delta y' + F_{z'}\, \delta z')\, dx.$$

Integrating by parts the last two terms of the integrand we have

$$\delta v = F|_{x=x_1} \delta x_1 + [F_{y'}\, \delta y]_{x=x_1} + [F_{z'}\, \delta z]_{x=x_1} +$$

$$+ \int\limits_{x_0}^{x_1} \left(\left(F_y - \frac{d}{dx}\, F_{y'}\right) \delta y + \left(F_z - \frac{d}{dx}\, F_{z'}\right) \delta z\right) dx.$$

Since v is now considered along extremals only

$$F_y - \frac{d}{dx}\, F_{y'} \equiv 0, \quad F_z - \frac{d}{dx}\, F_{z'} \equiv 0,$$

and consequently

$$\delta v = F_{x=x_1} \delta x_1 + [F_{y'}\, \delta y]_{x=x_1} + [F_{z'}\, \delta z]_{x=x_1}.$$

By the same argument as that given on p. 68, we obtain

$$\delta y\big|_{x=x_1} = \delta y_1 - y'(x_1)\,\delta x_1 \quad \text{and} \quad \delta z\big|_{x=x_1} = \delta z_1 - z'(x_1)\,\delta x_1,$$

and consequently

$$\delta v = [F - y'F_{y'} - z'F_{z'}]_{x=x_1}\,\delta x_1 + F_{y'}\big|_{x=x_1}\,\delta y_1 +$$
$$+ F_{z'}\big|_{x=x_1}\,\delta z_1 = 0.$$

If the variations δx_1, δy_1, δz_1 are independent, then it follows from the condition $\delta v = 0$ that

$$[F - y'F_{y'} - z'F_{z'}]_{x=x_1} = 0, \quad F_{y'}\big|_{x=x_1} = 0, \quad F_{z'}\big|_{x=x_1} = 0.$$

If the boundary point $B(x_1, y_1, z_1)$ can move along a certain curve $y_1 = \varphi(x_1)$, $z_1 = \psi(x_1)$, then $\delta y_1 = \varphi'(x_1)\,\delta x_1$ and $\delta z_1 = \psi'(x_1)\,\delta x_1$ and the condition $\delta v = 0$ or

$$[F - y'F_{y'} - z'F_{z'}]_{x=x_1}\,\delta x_1 + F_{y'}\big|_{x=x_1}\,\delta y_1 + F_{z'}\big|_{x=x_1}\,\delta z_1 = 0$$

turns into

$$[F + (\varphi' - y')F_{y'} + (\psi' - z')F_{z'}]_{x=x_1}\,\delta x_1 = 0,$$

and since δx_1 is arbitrary, we have

$$[F + (\varphi' - y')F_{y'} + (\psi' - z')F_{z'}]_{x=x_1} = 0.$$

This condition is called the *transversality condition* for the problem of determining the extrema of the functional

$$v = \int_{x_0}^{x_1} F(x, y, z, y', z')\,dx.$$

This transversality condition along with the equations $y_1 = \varphi(x_1)$, $z_1 = \psi(x_1)$ is not enough to evaluate all arbitrary constants in the general solution of a system of Euler equations.

If the boundary point $B(x_1, y_1, z_1)$ can move on a surface $z_1 = \varphi(x_1, y_1)$, then

$$\delta z_1 = \varphi'_{x_1}\,\delta x_1 + \varphi'_{y_1}\,\delta y_1,$$

where the variations δx_1 and δy_1 are arbitrary and independent. Consequently, the condition $\delta v = 0$ written in full

$$[F - y' F_{y'} - z' F_{z'}]_{x=x_1} \delta x_1 + F_{y'}|_{x=x_1} \delta y_1 + F_{z'}|_{x=x_1} \delta z_1 = 0$$

takes the form

$$[F - y' F_{y'} - z' F_{z'} + \varphi'_x F_{z'}]_{x=x_1} \delta x_1 +$$
$$+ [F_{y'} + F_{z'} \varphi'_y]_{x=x_1} \delta y_1 = 0.$$

Hence, since δx_1 and δy_1 are independent, we have

$$[F - y' F_{y'} + (\varphi'_x - z') F_{z'}]_{x=x_1} = 0, \quad [F_{y'} + F_{z'} \varphi'_y]_{x=x_1} = 0.$$

These two relations along with $z_1 = \varphi(x_1, y_1)$, are in general enough to evaluate two arbitrary constants of general solution of the system of Euler equations.

If the boundary point $A(x_0, y_0, z_0)$ is also variable, then by the same argument we can obtain similar conditions for this point.

If we consider the functional

$$v = \int_{x_0}^{x_1} F(x, y_1, y_2, \ldots, y_n, y'_1, y'_2, \ldots, y'_n) \, dx$$

with one variable boundary point $B(x_1, y_{11}, y_{21}, \ldots, y_{n1})$, then by a similar argument we find that the following relation holds at this point

$$\left(F - \sum_{i=1}^{n} y'_i F'_{y_i} \right)\Big|_{x=x_1} \delta x_1 + \sum_{i=1}^{n} F'_{y_i}\Big|_{x=x_1} \delta y_{i1} = 0.$$

EXAMPLE 1. *Find the transversality condition for the functional*

$$v = \int_{x_0}^{x_1} A(x, y, z) \sqrt{1 + y'^2 + z'^2} \, dx, \quad \text{where} \quad z_1 = \varphi(x_1, y_1).$$

Transversality conditions

$$[F - y' F_{y'} + (\varphi'_x - z') F_{z'}]_{x=x_1} = 0 \quad \text{and} \quad [F_{y'} + F_{z'} \varphi'_y]_{x=x_1} = 0$$

in this case, turn into $1 + \varphi'_x z' = 0$ and $y' + \varphi'_y z' = 0$ at $x = x_1$, or

$$\frac{1}{\varphi'_x} = \frac{y'}{\varphi'_y} = \frac{z'}{-1} \quad \text{at} \quad x = x_1.$$

This relation means that the tangent vector $\bar{l}(1, y', z')$ to the extremal at the point (x_1, y_1, z_1) and the normal vector $\overline{N}(\varphi'_x, \varphi'_y, -1)$ to the surface $z = \varphi(x, y)$ at the same point are parallel. Consequently, in this case the transversality condition means that the extremal should be orthogonal to the surface $z = \varphi(x, y)$.

EXAMPLE 2. *Find the extremum of distance between two surfaces*

$$z = \varphi(x, y) \quad and \quad z = \psi(x, y).$$

In other words, find the extrema of the integral

$$l = \int_{x_0}^{x_1} \sqrt{1 + y'^2 + z'^2}\ dx$$

with an additional condition that the coordinates of one boundary point satisfy the equation $z_0 = \varphi(x_0, y_0)$ and those of the other point satisfy the equation $z_1 = \psi(x_1, y_1)$.

Since the integrand depends only on y' and z', the extremals are straight lines (cf. Example 2, p. 42). And since the functional

$$\int_{x_0}^{x_1} \sqrt{1 + y'^2 + z'^2}\ dx$$

is a particular case of the functional

$$\int_{x_0}^{x_1} A(x, y, z) \sqrt{1 + y'^2 + z'^2}\ dx$$

considered in the previous example, it follows that the transversality conditions both at the point (x_0, y_0, z_0) and (x_1, y_1, z_1) turn into orthogonality conditions. Consequently, an extremum can occur only along such straight lines that are orthogonal to $z = \varphi(x, y)$ at the point (x_0, y_0, z_0) and to $z = \psi(x, y)$ at the point (x_1, y_1, z_1) (Fig. 24).

EXAMPLE 3. *Examine the extrema of the functional*

$$v = \int_0^{x_1} (y'^2 + z'^2 + 2yz)\, dx,$$

where $y(0) = 0$, $z(0) = 0$, *and the point* (x_1, y_1, z_1) *can move on the plane* $x = x_1$.

The system of Euler equations is $z'' - y = 0$, $y'' - z = 0$, whence $y^{\mathrm{IV}} - y = 0$, $y = C_1 \cosh x + C_2 \sinh x + C_3 \cos x + C_4 \sin x$, $z = y''$, $z = C_1 \cosh x + C_2 \sinh x = C_3 \cos x - C_4 \sin x$. From the conditions $y(0) = 0$ and $z(0) = 0$ we obtain $C_1 + C_3 = 0$ and $C_1 - C_3 = 0$, hence $C_1 = C_3 = 0$. The condition at the variable point

$$[F - y' F_{y'} - z' F_{z'}]_{x=x_1} \delta x_1 + F_{y'}|_{x=x_1} \delta y_1 + F_{z'}|_{x=x_1} \delta z_1 = 0$$

turns into

$$F_{y'}|_{x=x_1} = 0 \quad \text{and} \quad F_{z'}|_{x=x_1} = 0,$$

FIG. 24

for $\delta x_1 = 0$ and δy_1 and δz_1 are independent. In the present case $F_{y'} = 2y'$, $F_{z'} = 2z'$, and consequently

$$y'(x_1) = 0 \quad \text{and} \quad z'(x_1) = 0$$

or

$$C_2 \cosh x_1 + C_4 \cos x_1 = 0 \quad \text{and} \quad C_2 \cosh x_1 - C_4 \cos x_1 = 0.$$

If $\cos x_1 \neq 0$, then $C_2 = C_4 = 0$ so that an extremum can occur only along the straight line $y = 0$, $z = 0$. When $\cos x_1 = 0$, so that $x_1 = \frac{1}{2}\pi + n\pi$ where n is an integer, then it follows that $C_2 = 0$, and that C_4 is an arbitrary constant, so that

$$y = C_4 \sin x, \quad z = -C_4 \sin x.$$

It is readily verified that in this latter case we have $v = 0$ for arbitrary C_4.

3. Problems with movable boundaries for functionals of the form $\int\limits_{x_0}^{x_1} F(x, y, y', y'')\,dx$

Investigating the extrema of the functional

$$v\big(y(x)\big) = \int\limits_{x_0}^{x_1} F(x, y, y', y'')\,dx$$

in Section 4, Chapter I we assumed that the values of the function and its first derivative at the boundary point are given,

$$y(x_0) = y_0, \quad y'(x_0) = y_0', \quad y(x_1) = y_1, \quad y'(x_1) = y_1'.$$

If some of these values are variable, then the corresponding variational problem is called a *problem with variable or movable boundaries*.

If there is an extremum on a curve C for a problem with movable boundaries, then it is also extremum with respect to a more restricted class of neighbouring curves that have end points in common with the curve C and the same direction of tangents at these points. Consequently the curve C should satisfy the equations of Euler-Poisson (cf. p. 45),

$$F_y - \frac{d}{dx} F_{y'} + \frac{d^2}{dx^2} F_{y''} = 0.$$

The general solution of this fourth-order differential equation $y = y(x, C_1, C_2, C_3, C_4)$ involves four arbitrary constants. Four additional relations are required to determine these constants. These relations can be obtained from the fundamental necessary condition of an extremum $\delta v = 0$.

To make the calculation less cumbersome, let us suppose that the boundary values of y and y' at the point x_0 are fixed in advance $y(x_0) = y_0$, $y'(x_0) = y_0'$, and the other point is variable. In general, utilizing these boundary conditions $y(x_0) = y_0$, $y'(x_0) = y_0'$, we can evaluate two of the arbitrary constants. The remaining two can be determined from the condition $\delta v = 0$, where the functional is now considered, restricted to the solutions of the Euler-Poisson equation, for an extremum can occur only along these curves. The main linear part of an increment Δv of the functional can be calculated as in Section 1, Chapter II:

$$\Delta v = \int\limits_{x_0}^{x_1+\delta x_1} F(x, y+\delta y, y'+\delta y', y''+\delta y'')\,dx -$$

$$- \int\limits_{x_0}^{x_1} F(x, y, y', y'')\,dx$$

$$= \int\limits_{x_0}^{x_1+\delta x_1} F(x, y+\delta y, y'+\delta y', y''+\delta y'')\,dx +$$

$$+ \int\limits_{x_0}^{x_1} \big(F(x, y+\delta y, y'+\delta y', y''+\delta y'') -$$

$$- F(x, y, y', y'')\big)\,dx.$$

Applying the mean value theorem and using the continuity of the functions F and $y(x)$, $y'(x)$, $y''(x)$ we have

$$\Delta v = F(x, y, y'')\big|_{x=x_1}\,\delta x_1 +$$

$$+ \int\limits_{x_0}^{x_1} (F_y\,\delta y + F_{y'}\,\delta y' + F_{y''}\,\delta y'')\,dx + R,$$

where R is an infinitesimal of order higher than the maximum of the absolute values $|\delta x_1|$, $|\delta y_1|$, $|\delta y|$, $|\delta y'|$, and $|\delta y''|$. Consequently,

$$\delta v = F\big|_{x=x_1}\,\delta x_1 + \int\limits_{x_0}^{x_1} (F_y\,\delta y + F_{y'}\,\delta y' + F_{y''}\,\delta y'')\,dx.$$

Integrating by parts the second term of the integrand and doing the same twice with the third term, and then remembering that

$$\delta y|_{x=x_0} = 0, \quad \delta y'|_{x=x_0} = 0, \quad \text{and} \quad F_y - \frac{d}{dx}F_{y'} + \frac{d^2}{dx^2}F_{y''} \equiv 0,$$

we obtain

$$\delta v = \left[F\delta x_1 + F_{y'}\,\delta y + F_{y''}\,\delta y' - \frac{d}{dx}(F_{y''})\,\delta y\right]_{x=x_1}.$$

Making use of the relation $\delta y_1 = y'(x_1)\,\delta x_1 + [\delta y]_{x=x_1}$ (cf. p. 69) and applying this result also to $\delta y_1'$:

$$\delta y_1' = y''(x_1)\,\delta x_1 + [\delta y']_{x=x_1},$$

we obtain

$$\delta v = \left[F - y'F_{y'} - y''F_{y''} + y'\frac{d}{dx}(F_{y''})\right]_{x=x_1}\delta x_1 +$$

$$+ \left[F_{y'} - \frac{d}{dx}(F_{y''})\right]_{x=x_1}\delta y_1 + F_{y''}|_{x=x_1}\delta y_1'.$$

Consequently, the fundamental condition of an extremum $\delta v = 0$ takes the form

$$(\text{I}) \quad \left[F - y'F_{y'} - y''F_{y''} + y'\frac{d}{dx}(F_{y''})\right]_{x=x_1}\delta x_1 +$$

$$+ \left[F_{y'} - \frac{d}{dx}(F_{y''})\right]_{x=x_1}\delta y_1 + F_{y''}|_{x=x_1}\delta y_1' = 0.$$

If δx_1, δy_1 and $\delta y_1'$ are independent, then their coefficients should vanish at the point $x = x_1$. If there is some relation between them, for instance $y_1 = \varphi(x_1)$ and $y_1' = \psi(x_1)$, then $\delta y_1 = \varphi'(x_1)\,\delta x_1$ and $\delta y_1' = \psi'(x_1)\,\delta x_1$, and substituting these values into (I) we have

$$\left[F - y'F_{y'} - y''F_{y''} + y'\frac{d}{dx}(F_{y''}) + \left(F_{y'} - \frac{d}{dx}F_{y''}\right)\varphi' +\right.$$

$$\left. + F_{y''}\psi'\right]_{x=x_1}\delta x_1 = 0,$$

hence

$$\left[F - y' F_{y'} - y'' F_{y''} + y' \frac{d}{dx}(F_{y''}) + \left(F_{y'} - \frac{d}{dx} F_{y''} \right) \varphi' + \right.$$

$$\left. + F_{y''} \psi' \right]_{x=x_1} = 0.$$

This condition along with the equations $y_1 = \varphi(x_1)$ and $y_1' = \psi(x_1)$ in general are enough to determine the values of $x_1, y_1,$ and y_1'.

If x_1, y_1, y_1' are related through one equation $\varphi(x_1, y_1, y_1') = 0$, then two of the variations $\delta x_1, \delta y_1, \delta y_1'$ are arbitrary and the remaining one is given by the equation

$$\varphi_{x_1}' \delta x_1 + \varphi_{y_1}' \delta y_1 + \varphi_{y_1'}' \delta y_1' = 0,$$

for instance

$$\delta y_1' = - \frac{\varphi_{x_1}' \delta x_1 + \varphi_{y_1}' \delta y_1}{\varphi_{y_1'}'} \quad \text{provided} \quad \varphi_{y_1'}' \neq 0.$$

Substituting $\delta y_1'$ into (I) and remembering that the coefficients of the independent variations δx_1 and δy_1 must vanish, we obtain two equations at the point $x = x_1$, which in general, together with $\varphi(x_1, y_1, y_1') = 0$ enable us to evaluate x_1, y_1, y_1'. If the point $A(x_0, y_0)$ is variable, too, then the analogous conditions can be obtained for this point.

EXAMPLE. *Examine the extrema of the functional*

$$v = \int_0^1 (1 + y''^2) dx, \quad where \quad y(0) = 0, \quad y'(0) = 1, \quad y(1) = 1,$$

and $y'(1)$ *is variable.*
The Euler-Poisson equation

$$F_y - \frac{d}{dx} F_{y'} + \frac{d^2}{dx^2} F_{y''} = 0$$

in this case, is $y^{iv} = 0$, hence $y = C_1 + C_2 x + C_3 x^2 + C_4 x^3$. From the condition $y(0) = 0$, we have $C_1 = 0$, and from $y'(0) = 1$, we have $C_2 = 1$. From $y(1) = 1$, it follows that $C_3 + C_4 = 0$. Since $\delta x_1 = 0$

and $\delta y_1 = 0$, whereas $\delta y_1'$ is arbitrary, it follows that condition (I) reduces to

$$F_{y''}\big|_{x=x_1} = 0, \quad \text{or} \quad y''(1) = 0.$$

Since $y''(x) = 2C_3 + 6C_4 x$, for $x = 1$, we have $2C_3 + 6C_4 = 0$, and since $C_3 + C_4 = 0$, we obtain $C_3 = 0$, $C_4 = 0$. Therefore an extremum can occur only along the straight line $y = x$.

4. Extremals with cusps

So far, only such variational problems were considered where the unknown function $y = y(x)$ was assumed to be continuous, and have continuous first derivative. However, there are a number of problems, for which

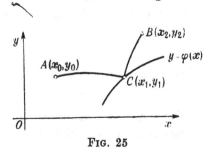

Fig. 25

this second assumption is not essential; moreover, there are such variational problems that have extrema only along extremals with cusps. For instance, the problems of reflection of extremals, or the problems of refraction of extremals are such, which are generalizations of the problems of reflection or refraction of light.

Problems with reflected extremals. Find the curve that makes a functional $v = \int_{x_0}^{x_2} F(x, y, y')\,dx$ an extremum and that passes through given points $A(x_0, y_0)$ and $B(x_2, y_2)$, where it is assumed that this curve should reach the point B after having been reflected by a given arc $y = \varphi(x)$ (Fig. 25).

It is evident that the point of reflection $C(x_1, y_1)$ may be a cusp of the extremal, and consequently the left-side derivative $y'(x_1 - 0)$ and the right-side derivative $y'(x_1 + 0)$ are not, in general, equal. It is therefore most convenient to express the functional $v(y(x))$ in the following form

$$v(y(x)) = \int_{x_0}^{x_1} F(x, y, y') dx + \int_{x_1}^{x_2} F(x, y, y') dx$$

where the derivative $y'(x)$ is assumed to be continuous on each of the intervals $x_0 \leqslant x \leqslant x_1$ and $x_1 \leqslant x \leqslant x_2$, so that we can rely on our previous results.

The fundamental condition for an extremum $\delta v = 0$ takes the form

$$\delta v = \delta \int_{x_0}^{x_1} F(x, y, y') dx + \delta \int_{x_1}^{x_2} F(x, y, y') dx = 0.$$

Since the point (x_1, y_1) can move along the curve $y = \varphi(x)$, it follows that the calculation of the variations

$$\delta \int_{x_0}^{x_1} F(x, y, y') dx \quad \text{and} \quad \delta \int_{x_1}^{x_2} F(x, y, y') dx$$

can be carried out as if in case of a problem with one variable boundary point, and we can use the results of Section 1, Chapter II (p. 69). It is evident that the curves AC and CB are extremals, so that along each of these arcs $y = y(x)$ is a solution of the Euler equation, for if we consider that one of these curves is already found, then in order to find the second curve, it is sufficient to find an extremum of the functional $\int_{x_0}^{x_1} F dx$ (or $\int_{x_1}^{x_2} F dx$), where the end points have been already fixed. Consequently, calculating the variation, we can assume that the functional varies only along the extremals with a bend point C. Thus

$$\delta \int_{x_0}^{x_1} F(x; y, y') \, dx = [F + (\varphi' - y') F_{y'}]_{x=x_1-0} \delta x_1$$

and

$$\delta \int_{x_1}^{x_2} F(x, y, y') \, dx = -[F + (\varphi' - y') F_{y'}]_{x=x_1+0} \delta x_1$$

(cf. p. 69), where the subscripts $x = x_1 - 0$ and $x = x_1 + 0$ mean that we should take the limit value of the expression in bracket as x tends to x_1 from the left side, i.e. being less than x_1, or from the right side, i.e. being greater than x_1, correspondingly. Since at the point of reflection only the derivative is discontinuous, it follows that in the first case the left-side derivative shall be taken at this cusp, and in the second case, the right-side derivative.

The condition $\delta v = 0$ takes the form

$$[F + (\varphi' - y') F_{y'}]_{x=x_1-0} \delta x_1 - [F + (\varphi' - y') F_{y'}]_{x=x_1+0} \delta x_1 = 0$$

or, since δx_1 varies arbitrarily,

$$[F + (\varphi' - y') F_{y'}]_{x=x_1-0} = [F + (\varphi' - y') F_{y'}]_{x=x_1+0},$$

or

$$F\big(x_1, y_1, y'(x_1-0)\big) + \big(\varphi'(x_1) - y'(x_1-0)\big) \times$$
$$\times F_{y'}\big(x_1, y_1, y'(x_1-0)\big) = F\big(x_1, y_1, y'(x_1+0)\big) +$$
$$+ \big(\varphi'(x_1) - y'(x_1+0)\big) F_{y'}\big(x_1, y_1, y'(x_1+0)\big).$$

This reflection condition has a particularly simple form for functionals of the form

$$v = \int_{x_0}^{x_1} A(x, y) \sqrt{1 + y'^2} \, dx,$$

namely

$$A(x_1, y_1)\left[\sqrt{1+y'^2} + \frac{(\varphi' - y')y'}{\sqrt{1+y'^2}}\right]_{x=x_1-0}$$

$$= A(x_1, y_1)\left[\sqrt{1+y'^2} + \frac{(\varphi' - y')y'}{\sqrt{1+y'^2}}\right]_{x=x_1+0}$$

or reducing and dividing by $A(x_1, y_1)$, assuming that $A(x_1, y_1) \neq 0$, we obtain

$$\frac{1+\varphi' y'}{\sqrt{1+y'^2}}\bigg|_{x=x_1-0} = \frac{1+\varphi' y'}{\sqrt{1+y'^2}}\bigg|_{x=x_1+0}$$

If the angle between the tangent to the curve $y = \varphi(x)$ and the x-axis is denoted by α, and the angles of inclina-

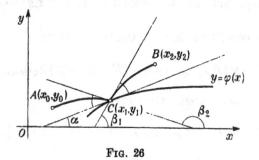

FIG. 26

tion with respect to the x-axis of the left-side tangent and right-side tangent at the reflection point are denoted by β_1 and β_2 respectively (Fig. 26), then

$$y'(x_1 - 0) = \tan\beta_1, \quad y'(x_1 + 0) = \tan\beta_2, \quad \varphi'(x_1) = \tan\alpha.$$

The condition at the point of reflection takes the form

$$\frac{1+\tan\alpha\tan\beta_1}{-\sec\beta_1} = \frac{1+\tan\alpha\tan\beta_2}{\sec\beta_2},$$

which when multiplied by $\cos \alpha$ reduces to

$$-\cos(\alpha - \beta_1) = \cos(\alpha - \beta_2).$$

Hence the angle of incidence is equal to the angle of reflection.

If a point is moving in a medium with velocity $v(x, y)$, then the time t in which it travels from $A(x_0, y_0)$ to $B(x_1, y_1)$ is given by the integral

$$\int_{x_0}^{x_1} \frac{\sqrt{1+y'^2}}{v(x, y)} \, dx,$$

which belongs to the above considered class of functionals

$$\int_{x_0}^{x_1} A(x, y) \sqrt{1+y'^2} \, dx,$$

and therefore, even if the velocity is given by an arbitrary function $v(x, y)$ the angle of incidence at the reflection point is equal to the angle of reflection.

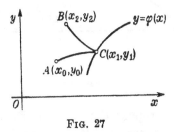

FIG. 27

If the points A and B were situated in some other way, for instance, like at Fig. 27, then since $y = y(x)$ would not be a univalent function we should investigate this problem in a parametric representation.

Refraction of extremals. Let us assume that there is a curve of discontinuity $y = \varphi(x)$ of the integrand F of the functional

$$v = \int_{x_0}^{x_2} F(x, y, y')\,dx$$

in the domain of F, and that the limit points A and B lie on the opposite sides of this curve of discontinuity (Fig. 28).

Let us represent the functional v in the form

$$v = \int_{x_0}^{x_1} F_1(x, y, y')\,dx + \int_{x_1}^{x_2} F_2(x, y, y')\,dx$$

where $F_1(x, y, y') = F(x, y, y')$ on one side of the curve of discontinuity, and $F_2(x, y, y') = F(x, y, y')$ on the other side.

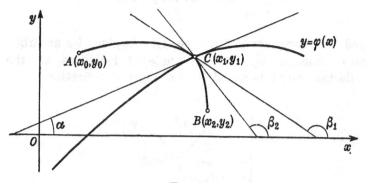

FIG. 28

Further, we suppose that F_1 and F_2 have third-order derivatives. It is reasonable to accept that the point of intersection C of a curve giving extremum with the curve of discontinuity will be a cusp. The paths AC and CB are, of course, extremals, for again considering one of these paths fixed and varying the other we obtain a problem with fixed endpoints. Therefore, the class of comparison curves can be restricted to all curves with corners consisting of two extremal paths, and since the cusp $C(x_1, y_1)$ is movable, moving along the curve $y = \varphi(x)$, the variation is (cf. p. 69):

$$\delta v = \delta \int_{x_0}^{x_1} F_1(x, y, y')\, dx + \delta \int_{x_1}^{x_2} F_2(x, y, y')\, dx$$

$$= [F_1 + (\varphi' - y')F_{1y'}]_{x=x_1-0}\, \delta x_1 -$$
$$- [F_2 + (\varphi' - y')F_{2y'}]_{x=x_1+0}\, \delta x_1.$$

Consequently the fundamental necessary condition for an extremum $\delta v = 0$ reduces to the equation

$$[F_1 + (\varphi' - y')F_{1y'}]_{x=x_1-0} = [F_2 + (\varphi' - y')F_{2y'}]_{x=x_1+0}.$$

Since at the refraction point only y' may fail to be continuous, it follows that this latter refraction condition can be written down as

$$F_1\big(x_1, y_1, y'(x_1-0)\big) +$$
$$+ \big(\varphi'(x_1) - y'(x_1-0)\big)F_{1y'}\big(x_1, y_1, y'(x_1-0)\big)$$
$$= F_2\big(x_1, y_1, y'(x_1+0)\big) +$$
$$+ \big(\varphi'(x_1) - y'(x_1+0)\big)F_{2y'}\big(x_1, y_1, y'(x_1+0)\big).$$

This refraction condition along with the equation $y_1 = \varphi(x_1)$ enables one to evaluate the coordinates of the point C.

If, in particular, the functional v is of the form

$$\int_{x_0}^{x_2} A(x, y)\sqrt{1+y'^2}\, dx$$

$$= \int_{x_0}^{x_1} A_1(x, y)\sqrt{1+y'^2}\, dx + \int_{x_1}^{x_2} A_2(x, y)\sqrt{1+y'^2}\, dx$$

then the refraction condition takes the form

$$A_1(x, y)\frac{1+\varphi' y'}{\sqrt{1+y'^2}}\Bigg|_{x=x_1-0} = A_2(x, y)\frac{1+\varphi' y'}{\sqrt{1+y'^2}}\Bigg|_{x=x_1+0},$$

or, performing reduction with the same notation as that on p. 86

$$y'(x_1 - 0) = \tan \beta_1, \quad y'(x_1 + 0) = \tan \beta_2,$$

$$\varphi'(x_1) = \tan \alpha,$$

and multiplying by $\cos \alpha$ we have

$$\frac{\cos(\alpha - \beta_1)}{\cos(\alpha - \beta_2)} = \frac{A_2(x_1, y_1)}{A_1(x_1, y_1)}$$

or

$$\frac{\sin\left(\frac{1}{2}\pi - (\alpha - \beta_1)\right)}{\sin\left(\frac{1}{2}\pi - (\alpha - \beta_2)\right)} = \frac{A_2(x_1, y_1)}{A_1(x_1, y_1)},$$

which is a generalization of the well known principle of refraction of light: the sine of the angle of incidence and the angle of refraction are in the same ratio as the corresponding velocities

$$v_1(x, y) = \frac{1}{A_1(x, y)} \quad \text{and} \quad v_2(x, y) = \frac{1}{A_2(x, y)}$$

(cf. p. 86) in those media on the boundary of which the refraction occurs.

We should not be misled by these examples to the idea that the only problems where extremals with corners occur are reflection or refraction problems. Extrema may occur along extremals with corners even when the integrand F of the functional

$$v = \int_{x_0}^{x_1} F(x, y, y') dx$$

has all third derivatives, and all the admissible curves are subject only to the assumption that they should pass through two fixed end points A and B, without any further restrictions.

Let us investigate the functional

$$v = \int\limits_0^2 y'^2 (1 - y')^2 dx, \quad y(0) = 0, \quad y(2) = 1.$$

Since the integrand is positive, it follows that $v \geqslant 0$ and consequently, if there is any curve along which $v = 0$, then it surely makes v an absolute minimum, i.e. such curve gives the least value of v among all values that are taken on by v along the admissible curves. It is easy to see that the functional v vanishes along a polygonal curve

$$y = x \quad \text{for} \quad 0 \leqslant x \leqslant 1,$$

$$y = 1 \quad \text{for} \quad 1 \leqslant x \leqslant 2$$

(Fig. 29), for the integrand vanishes along this curve. Consequently, this polygonal curve makes v an absolute minimum.

There are also absolute minima of the functional v along the polygonal curves shown in Fig. 31. On the other

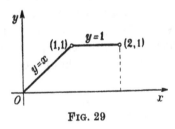

Fig. 29

hand, it is easy to see that along all smooth curves the values of the functional v are strictly positive even though these values approach zero arbitrarily close. In fact, the integrand vanishes only when $y = x + C_1$ or $y = C_2$. On the other hand the curves that consist of segments of straight lines and pass through the points $A(0, 0)$ and $B(2, 1)$ can be only polygonal curves. However, taking any such polygonal curve we can make it smooth by replacing only arbitrarily small parts of it contained in arbitrarily small neighbourhoods of its corner points, and the value of the functional along a curve so obtained will be arbitrarily close to the value of this functional taken on along that polygonal curve itself. Therefore $v = 0$ is exactly the greatest lower bound of the values of v taken on along smooth curves. This greatest lower bound is not attained along any smooth curve; it is attained only along piecewise smooth curves.

We now find the conditions that shall be satisfied by solutions with corners of a problem of extrema of the functional

$$v\big(y(x)\big) = \int\limits_{x_0}^{x_2} F(x, y, y')\, dx.$$

It is evident that the smooth pieces of arc of which an extremal with corners consists should coincide with pieces of arc of integral curves of Euler's equation. By fixing all smooth pieces of arc of which a given extremal consists save one, and varying this one smooth piece of arc we have a simple problem with fixed end points. Therefore this piece of arc should lie on an extremal.

For the sake of clarity we assume that the extremal to be found has only one cusp [1], and we will try to find some condition that should be satisfied at this bend point,

$$v = \int\limits_{x_0}^{x_2} F(x, y, y')\, dx = \int\limits_{x_0}^{x_1} F(x, y, y')\, dx +$$

$$+ \int\limits_{x_1}^{x_2} F(x, y, y')\, dx,$$

where x_1 is the x-coordinate of the cusp (Fig. 30). Since the curves AC and CB are integral curves of the

---

[1] If there are more of them, the following argument applies to each of them separately.

Euler equation, and since C varies arbitrarily, it follows from Section 1 Chapter II that

$$\delta v = (F - y' F_{y'})|_{x=x_1-0}\delta x_1 + F_{y'}|_{x=x_1-0}\,\delta y_1 -$$
$$- (F - y' F_{y'})|_{x=x_1+0}\delta x_1 - F_{y'}|_{x=x_1+0}\,\delta y_1 = 0,$$

hence

$$(F - y' F_{y'})|_{x=x_1-0}\delta x_1 + F_{y'}|_{x=x_1-0}\delta y_1$$
$$= (F - y' F_{y'})|_{x=x_1+0}\delta x_1 + F_{y'}|_{x=x_1+0}\delta y_1$$

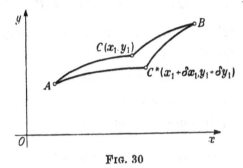

FIG. 30

or, since δx_1 and δy_1 are independent, we have

$$(F - y' F_{y'})|_{x=x_1-0} = (F - y' F_{y'})|_{x=x_1+0},$$

hence

$$F_{y'}|_{x=x_1-0} = F_{y'}|_{x=x_1+0}.$$

These conditions along with the continuity conditions of the cusp are enough to determine the coordinates of this cusp.

EXAMPLE 1. *Find the extremals with corners, if any, for the functional*

$$v = \int\limits_0^a (y'^2 - y^2)\,dx.$$

Let us write down the second condition that should be satisfied at a cusp, $F_{y'}|_{x=x_1-0} = F_{y'}|_{x=x_1+0}$, or, in the present case, $2y'(x_1 - 0) = 2y'(x_1 + 0)$, and consequently $y'(x_1 - 0)$

$= y'(x_1 + 0)$, i. e. the derivative y' is continuous at the point x_1, and therefore there are no cusps at all. Consequently in this case an extremum can occur only along a smooth curve.

EXAMPLE 2. *Find extremals with corners for the functional*

$$v = \int_{x_0}^{x_2} y'^2 (1 - y')^2 dx.$$

Since the integrand depends only on y', it follows that the extremals are straight lines $y = Cx + C^*$ (cf. p. 34). The conditions at a corner point take the form

$$- y'^2 (1 - y')(1 - 3y')|_{x = x_1 - 0} = - y'^2 (1 - y')(1 - 3y')|_{x = x_1 + 0}$$

or

$$2y'(1 - y')(1 - 2y')|_{x = x_1 - 0} = 2y'(1 - y')(1 - 2y')|_{x = x_1 + 0}.$$

Apart from the trivial case when

$$y'(x_1 - 0) = y'(x_1 + 0),$$

these conditions are satisfied when

$$y'(x_1 - 0) = 0 \quad \text{and} \quad y'(x_1 + 0) = 1$$

or

$$y'(x_1 - 0) = 1 \quad \text{and} \quad y'(x_1 + 0) = 0.$$

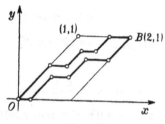

FIG. 31

Consequently, extremals with corners may consist only of some intervals of straight lines belonging to the family $y = C_1$ and $y = x + C_2$ (Fig. 31).

5. One-sided variations

For some variational problems of finding extrema of a functional $v(y(x))$, the class of admissible curves is sometimes subject to a restriction that they must pass through a certain domain R bounded by a certain curve $\Phi(x, y) = 0$ (Fig. 32). In such cases a curve C that gives an extremum should either pass entirely outside the boundary of R, and therefore it should be an extremal, for then the existence of R is no source of influence on the behaviour of v in some neighbourhood of the curve C and the arguments of Chapter I apply, or the curve C consists of a path lying outside the border of R and

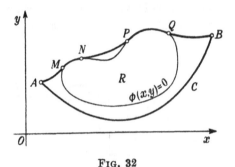

FIG. 32

a path leading along this border. In this latter case a new situation arises, at the part of the curve along the border of R. Only one-sided variations are possible there, for there is no admissible curve that touches the interior of R. Those parts of the curve C that lie outside R should as previously be extremals, for if we vary the curve C varying it only on such parts that admit of two-sided variation, then the existence of R does not have any influence on variation of $y(x)$, and the conclusions of Chapter I are valid.

Therefore in the case considered, an extremum can occur only along such curves that consist in part of extremals and in part of the boundary of R, and consequently in order to determine a curve giving an extremum some conditions are required at the points where extremals join with a piece of boundary of R. These conditions shall determine these points. In a case such as shown in Fig. 33, we shall derive these conditions at the points M, N, P and Q. For instance, let us find such condition at the point M. The conditions at the remaining points obtain likewise.

Calculating the variation δv of the functional

$$v = \int_{x_0}^{x_1} F(x, y, y') dx = \int_{x_0}^{x^*} F(x, y, y') dx + \int_{x^*}^{x_1} F(x, y, y') dx$$

we can assume that this variation is determined in full by a displacement of the point $M(x, y)$ on the curve

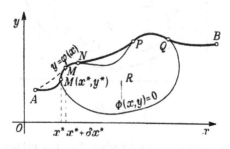

FIG. 33

$\Phi(x, y) = 0$, i.e. we can assume that for each position of the point M on the curve $\Phi(x, y) = 0$ the arc AM is always an extremal and the part $MNPQB$ remains fixed. The functional

$$v_1 = \int_{x_0}^{x^*} F(x, y, y') dx$$

has a movable upper limit, which can move along the boundary of R given by the equation $\Phi(x, y) = 0$. Written explicitly with respect to y, in a neighbourhood of the point M this equation is $y = \varphi(x)$.

Consequently, by Section 1 Chapter II,

$$\delta v_1 = [F + (\varphi' - y') F_{y'}]_{x=x^*} \, \delta x^*.$$

The functional

$$v_2 = \int\limits_{x^*}^{x_1} F(x, y, y')\, dx$$

has also a variable limit point, (x^*, y^*), but at the neighbourhood of this point the curve $y = \varphi(x)$ does not vary. Consequently, the whole change of the functional v_2 due to a displacement of the point (x^*, y^*) to $(x^* + \delta x^*, y^* + \delta y^*)$ reduces to that which is due to the change of the lower limit of integration

$$\Delta v_2 = \int\limits_{x^* + \delta x^*}^{x_1} F(x, y, y')\, dx - \int\limits_{x^*}^{x_1} F(x, y, y')\, dx$$

$$= -\int\limits_{x^*}^{x^* + \delta x^*} F(x, y, y')\, dx = -\int\limits_{x^*}^{x^* + \delta x^*} F\big(x, \varphi(x), \varphi'(x)\big)\, dx,$$

for $y = \varphi(x)$ on the interval $(x^*, x^* + \delta x^*)$.

Applying the mean value theorem and using the continuity of the function F we have

$$\Delta v_2 = -F\big(x, \varphi(x), \varphi'(x)\big)\big|_{x=x^*}\, \delta x^* + \beta\, \delta x^*,$$

where $\beta \to 0$ as $\delta x^* \to 0$. Consequently,

$$\delta v_2 = -F\big(x, \varphi(x), \varphi'(x)\big)\big|_{x=x^*}\, \delta x^*,$$

$$\delta v = \delta v_1 + \delta v_2$$

$$= [F(x, y, y') + (\varphi' - y')F_{y'}(x, y, y')]_{x=x^*}\, \delta x^* -$$
$$- F(x, y, \varphi')\big|_{x=x^*}\, \delta x^*$$

$$= [F(x, y, y') - F(x, y, \varphi') - (y' - \varphi')F_{y'}(x, y, y')]_{x=x^*}\, \delta x^*,$$

for $y(x^*) = \varphi(x^*)$.

Since δx^* is arbitrary, the necessary condition of an extremum $\delta v = 0$ takes the form

$$[F(x, y, y') - F(x, y, \varphi') - (y' - \varphi')F_{y'}(x, y, y')]_{x=x^*} = 0.$$

Applying the mean value theorem we find that

$$(y' - \varphi')[F_{y'}(x, y, q) - F_{y'}(x, y, y')]_{x=x^*} = 0,$$

where q is some value between $\varphi'(x^*)$ and $y'(x^*)$. Applying the mean value theorem once again we have

$$(y'-\varphi')(q-y')F_{y'y'}(x, y, q^*)_{x=x^*} = 0,$$

where q^* is some value between q and $y'(x^*)$. Let us assume further that $F_{y'y'}(x, y, q^*) \neq 0$. This assumption will prove later to be an essential one for a number of variational problems (cf. Chapter III). With this assumption the condition at the point M turns into $y'(x^*) = = \varphi'(x^*)$, for $q = y'$ only when $y'(x^*) = \varphi'(x^*)$, as q is always between $y'(x^*)$ and $\varphi'(x^*)$.

Therefore, at the point M the extremal arc AM and the boundary arc MN have the common tangent, which is a left-hand tangent of the curve $y = y(x)$, and the right-hand tangent of the curve $y = \varphi(x)$. Consequently, an extremal is always tangent to the boundary of the domain R at a point M.

6. Mixed problems

We sometimes have to examine the extrema of the functional

$$v = \int_{x_0}^{x_1} F(x, y, y')\,dx + \Phi(x_0, y_0, x_1, y_1),$$

where the coordinates of end points x_0, y_0, x_1, y_1 may be subject to further conditions, for instance $y_0 = \varphi(x_0)$ and $y_1 = \psi(x_1)$, or

$$v = \iint_D F\left(x, y, z, \frac{\partial z}{\partial x}, \frac{\partial z}{\partial y}\right) dx\,dy + \int_C \Phi\,ds,$$

where C is a variable contour curve encircling the domain of integration D, and ds is an element of arc length, or

$$v = \iiint_W F\left(x, y, z, u, \frac{\partial u}{\partial x}, \frac{\partial u}{\partial y}, \frac{\partial u}{\partial z}\right) dx\,dy\,dz + \iint_S \Phi\,dS,$$

where S is a variable boundary surface of the integration domain W. Such problems are called *mixed variational problems*, for one has to examine the extrema of a sum of two functionals of different kinds, which are related in some way to each other.

The method of solution is the same as for solution of the fundamental variational problems, but the boundary conditions are more complicated. For instance, consider the mixed problem of finding the extrema of the functional

$$v = \int_{x_0}^{x_1} F(x, y, y')\, dx + \Phi(x_0, y_0, x_1, y_1).$$

It is evident that an extremum can occur only along the solutions of Euler equation

$$F_y - \frac{d}{dx} F_{y'} = 0,$$

for if a curve C makes the functional v have an extremum for a problem with variable end points, then such curve will also do the same with respect to a more restricted class of admissible curves the end points of which coincide with the end points of the curve C, and along such curves the functional

$$v = \int_{x_0}^{x_1} F(x, y, y')\, dx + \Phi(x_0, y_0, x_1, y_1)$$

differs from the functional

$$\int_{x_0}^{x_1} F(x, y, y')\, dx$$

only by the constant term $\Phi(x_0, y_0, x_1, y_1)$, which has no influence at all on the extremal properties of the functional.

Let us calculate the variation δv, assuming that neighbouring curves are always extremals (cf. p. 68-69)

$$\delta v = [(F - y' F_{y'}) \delta x_1 + F_{y'} \delta y_1]_{x=x_1} -$$

$$- [(F - y' F_{y'}) \delta x_0 + F_{y'} \delta y_0]_{x=x_0} +$$

$$+ \frac{\partial \Phi}{\partial x_0} \delta x_0 + \frac{\partial \Phi}{\partial y_0} \delta y_0 + \frac{\partial \Phi}{\partial x_1} \delta x_1 + \frac{\partial \Phi}{\partial y_1} \delta y_1.$$

If the boundary points (x_0, y_0) and (x_1, y_1) can move along prescribed curves $y_0 = \varphi(x_0)$ and $y_1 = \psi(x_1)$, then $\delta y_0 = \varphi'(x_0) \delta x_0$ and $\delta y_1 = \psi'(x_1) \delta x_1$, and the necessary condition of an extremum takes the form

$$\left[F + F_{y'}(\psi' - y') + \frac{\partial \Phi}{\partial x_1} + \frac{\partial \Phi}{\partial y_1} \psi' \right]_{x=x_1} \delta x_1 -$$

$$- \left[F + F_{y'}(\varphi' - y') - \frac{\partial \Phi}{\partial x_0} - \frac{\partial \Phi}{\partial y_0} \varphi' \right]_{x=x_0} \delta x_0 = 0.$$

Since δx_0 and δx_1 are independent, we finally obtain

$$\left[F + F_{y'}(\psi' - y') + \frac{\partial \Phi}{\partial x_1} + \frac{\partial \Phi}{\partial y_1} \psi' \right]_{x=x_1} = 0$$

and

$$\left[F + F_{y'}(\varphi' - y') - \frac{\partial \Phi}{\partial x_0} - \frac{\partial \Phi}{\partial y_0} \varphi' \right]_{x=x_0} = 0.$$

To calculate the variation for a mixed problem containing multiple integrals, we have to calculate the variation of a multiple integral with variable boundary.

Problems

1. Find a solution with one corner for the problem of minima of the functional

$$v(y(x)) = \int_0^4 (y' - 1)^2 (y' + 1)^2 \, dx, \quad y(0) = 0, \quad y(4) = 2.$$

2. Are there any solutions with cusps for the problem of extrema of the functional

$$v(y(x)) = \int_{x_0}^{x_1} (y'^2 + 2xy - y^2)\,dx, \quad y(x_0) = y_0, \quad y(x_1) = y_1 ?$$

3. Are there any solutions with cusps for the problem of extrema of the functional

$$v(y(x)) = \int_0^{x_1} y'^3\,dx, \quad y(0) = 0, \quad y(x_1) = y_1 ?$$

4. Are there any solutions with cusps for the problem of extrema of the functional

$$v(y(x)) = \int_0^{x_1} (y'^4 - 6y'^2)\,dx, \quad y(0) = 0, \quad y(x_1) = y_1 ?$$

5. Find the transversality condition for the functional

$$v(y(x)) = \int_{x_0}^{x_1} A(x, y)\,e^{\arctan y'}\sqrt{1 + y'^2}\,dx, \quad A(x, y) \neq 0.$$

6. Using the fundamental necessary condition for an extremum $\delta v = 0$, find a function that can make the functional

$$v(y(x)) = \int_0^1 (y''^2 - 2xy)\,dx$$

an extremum. It is assumed that

$$y(0) = y'(0) = 0, \quad y(1) = \frac{1}{120}, \quad \text{and } y'(1) \text{ is arbitrary.}$$

7. Find the curves that can make the functional

$$v(y(x)) = \int_0^{10} y'^3\,dx, \quad y(0) = 0, \quad y(10) = 0$$

an extremum, subject to the additional condition that no admissible curve can pass through the interior of the circle

$$(x - 5)^2 + y^2 = 9.$$

8. Find the function that can make the functional

$$v(y(x)) = \int_0^{\pi/4} (y^2 - y'^2)\,dx, \quad y(0) = 0$$

an extremum, if it is assumed that the other endpoint can slide along the straight line $x = \pi/4$.

9. Making use of the fundamental necessary condition $\delta v = 0$ only, find the curve that can make the functional

$$v(y(x)) = \int\limits_0^{x_1} \frac{\sqrt{1+y'^2}}{y} \, dx, \quad y(0) = 0$$

an extremum, where the remaining end point (x_1, y_1) can move along the circumference of the circle $(x-9)^2 + y^2 = 9$.

10. Starting from the point $A(x_0, y_0)$ not lying on the curve $y = \varphi(x)$ we have to reach the point $C(x_2, y_2)$ lying on the curve $y = \varphi(x)$ in the shortest time. Outside the curve $y = \varphi(x)$ the velocity is constant and its value is v_1. The velocity with which we move along the curve $y = \varphi(x)$ is also constant, its value is v_2, and $v_2 > v_1$.

CHAPTER III

SUFFICIENCY CONDITIONS FOR AN EXTREMUM

1. Fields of extremals

If to each point of a domain D on the plane Oxy there exists exactly one curve that belongs to a family of curves $y = y(x, C)$, then this family of curves in the domain D is called a *field*, or more exactly a *proper field*. The slope $p(x, y)$ of the tangent to a curve $y = y(x, C)$ at a point (x, y) is called the *slope of the field* at the point (x, y).

For instance, the parallel straight lines $y = x + C$

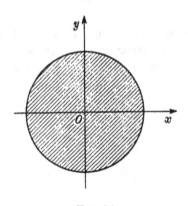

Fig. 34

form inside the circle $x^2 + y^2 = 1$ (Fig. 34) a field. The slope of this field is $p(x, y) = 1$. On the other hand, the family of parabolas $y = (x - a)^2 - 1$ (Fig. 35) is not a field inside the same circle, for inside this circle some of the parabolas intersect others.

If all the curves that belong to a family of curves

$y = y(x, C)$ pass through a certain point (x_0, y_0), i.e. if they form a pencil of curves, then of course, they do not form a proper field in the domain D, whenever the centre of this pencil belongs to D. However, if the curves

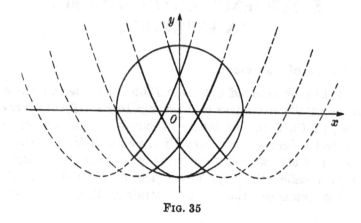

FIG. 35

of the pencil cover the whole domain D and never intersect with one another in this domain except at the centre of the pencil, i.e. if the conditions imposed on

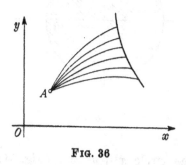

FIG. 36

fields hold at all points different from the centre of the pencil, then we say, too, that the family $y = y(x, C)$ is a field, but to distinguish such field from proper fields we call them *central fields* (Fig. 36).

For instance, the pencil of sine curves $y = C\sin x$ forms a central field in a sufficiently small neighbourhood of the interval $0 \leqslant x \leqslant a$ of x-axis, $a < \pi$ (Fig. 37). This same pencil of sine curves forms a proper field in a sufficiently small neighbourhood of the interval $\delta \leqslant x \leqslant a$ of x-axis, where $\delta > 0$, $a < \pi$ (Fig. 37).

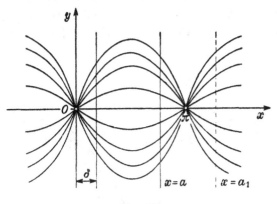

FIG. 37

If a proper or central field is generated by a family of extremals for some variational problem, then it is called a *field of extremals*.

The notion of a field of extremals can be extended almost without change to spaces of arbitrary dimension. A family $y_i = y_i(x, C_1, \ldots, C_n)$ $(i = 1, 2, \ldots, n)$ is a *field* in a domain D of the space $x, y_1, \ldots, y_n$, if through each point of this $n+1$ dimensional domain D there passes one and only one curve of the family $y_i = y_i(x, C_1, \ldots, C_n)$. The partial derivatives of the functions $y_i(x, C_1, C_2, \ldots, C_n)$ with respect to x taken at the point $(x, y_1, y_2, \ldots, y_n)$ are called the *slope functions*, and are denoted by $p_i(x, y_1, y_2, \ldots, y_n)$. To calculate $p_i(x, y_1, y_2, \ldots, y_n)$ one has to take $y_i(x, C_1, C_2, \ldots, C_n)$ and replace every constant, $C_1, C_2, \ldots, C_n$, by a formula that expresses it in terms of $x, y_1, y_2, \ldots, y_n$. The central field can be defined likewise.

Suppose that the curve $y = y(x)$ is an extremal curve for the variational problem of finding the extrema of the simplest functional

$$v(y(x)) = \int_{x_0}^{x_1} F(x, y, y')\,dx,$$

where the end points $A(x_0, y_0)$ and $B(x_1, y_1)$ are fixed. We say that the extremal curve $y = y(x)$ is admissible in a field of extremals, if there is a family of extremals $y = y(x, C)$ which is a field and which for a particular value of $C = C_0$, turns into the extremal $y = y(x)$

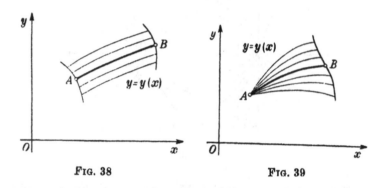

FIG. 38   FIG. 39

not lying on the boundary of the domain D in which the family $y = y(x, C)$ is a field (Fig. 38). If the pencil of extremals with the centre at the point $A(x_0, y_0)$ is a field in some neighbourhood of an extremal $y = y(x)$ passing through this point, then of course, there is a central field that admits of the given extremal $y = y(x)$. As a parameter of the family we may take the slope of the tangent to a curve of the pencil at the point $A(x_0, y_0)$ (Fig. 39).

EXAMPLE 1. *Given the functional*

$$\int_0^a (y'^2 - y^2)\,dx,$$

find a central field of extremals admitting the extremal arc $y = 0$, joining the points $(0, 0)$ and $(a, 0)$, where $0 < a < \pi$.

The general solution of the Euler equation $y'' + y = 0$ (cf. p. 31, Example 1) is $y = C_1 \cos x + C_2 \sin x$. From the condition that the required extremal should pass through the point $(0, 0)$, it follows that $C_1 = 0$, $y = C_2 \sin x$, so that in an interval $0 \leqslant x \leqslant a$, $a < \pi$, the curves of this pencil form a central field including, when $C_2 = 0$, the extremal $y = 0$. The parameter C_2 of this family is the value of the derivative y'_x at the point $(0, 0)$. If in this same problem $a \geqslant \pi$, then the family $y = C_2 \sin x$ is not a field (see Fig. 37).

As is known, any two infinitesimally close curves of the family of curves $F(x, y, C) = 0$ intersect at a point lying on the C-discriminant locus determined by the equations

$$F(x, y, C) = 0, \qquad \frac{\partial F}{\partial C} = 0.$$

We recall that, in particular, a C-discriminant locus consists of the envelope of the family and the locus of the multiple point of a curve of the family. If $F(x, y, C) = 0$ is an equation of a pencil of curves, then also the centre of the pencil lies on the C-discriminant locus. Therefore, if we take a pencil of extremals $y = y(x, C)$ passing through a point (x_0, y_0) and determine its C-discriminant locus $\Phi(x, y) = 0$, then any two neighbouring curves of the family $y = y(x, C)$ intersect at a point lying near the curve $\Phi(x, y) = 0$. In particular, any two curves of the family which are neighbouring to a given extremal $y = y(x)$ which passes through the points $A(x_0, y_0)$ and $B(x_1, y_1)$, intersect near that point where the curve $y = y(x)$ touches the C-discriminant locus (see Fig. 40, where the C-discriminant locus is marked with thick line). If the arc AB of an extremal $y = y(x)$ has no point in common with the C-discriminant locus of a pencil of extremals admitting this given extremal, then those extremals of the pencil which are sufficiently close to the arc AB do not intersect, i.e. they form a central field including the arc AB in some neighbourhood of this arc (Fig. 41).

If the arc AB of an extremal $y = y(x)$ has a point in common A^*, distinct from A, with the C-discriminant locus of the pencil $y = y(x, C)$, then the curves neighbouring to $y = y(x)$ may intersect near the point A^* with one another, or with the curve $y = y(x)$. Therefore in general, they do not form a field (Fig. 40). A^* is called the point *conjugate* to A.

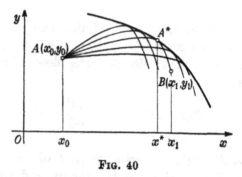

Fig. 40

This result can be stated as follows. To form a central field of extremals with centre at a point A admitting an arc AB of an extremal curve, it is sufficient that the point A^*

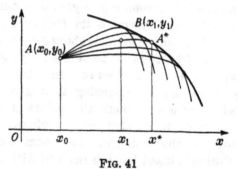

Fig. 41

conjugate to the point A does not lie on the arc AB. This condition which enables us to form a field of extremals admitting a prescribed extremal is called *Jacobi's condition*.

It is not difficult to reformulate this condition analytically. Let $y = y(x, C)$ be the equation of a pencil of extremals with centre at a point A, where the parameter C may be thought of as the slope y' of an extremal of the pencil at the point A. The C-discriminant locus is determined by the equations

$$y = y(x, C), \qquad \frac{\partial y(x, C)}{\partial C} = 0.$$

The derivative $\partial y(x, C)/\partial C$ taken along an arbitrary curve of the family turns into a function of x only. This function will be designated by u, $u = \partial y(x, C)/\partial C$, where C is fixed. Thus $u'_x = \partial^2 y(x, C)/\partial C \partial x$. The function $y = y(x, C)$ being a solution of the Euler equation, we have

$$F_y\big(x, y(x, C), y'_x(x, C)\big) - \frac{d}{dx}\, F_{y'}\big(x, y(x, C), y'_x(x, C)\big) \equiv 0.$$

Differentiating this identity with respect to C and setting $\partial y(x, C)/\partial C = u$, we obtain

$$F_{yy}u + F_{yy'}u' - \frac{d}{dx}\left(F_{yy'}u + F_{y'y'}u'\right) = u$$

or

$$\left(F_{yy} - \frac{d}{dx}\, F_{yy'}\right)u - \frac{d}{dx}\left(F_{y'y'}u'\right) = 0.$$

The functions $F_{yy}(x, y, y')$, $F_{yy'}(x, y, y')$, $F_{y'y'}(x, y, y')$ are known functions of x, for their second argument y is given by a solution $y = y(x, C)$ of the Euler equation, where $C = C_0$, which corresponds to the given extremal AB. This linear homogeneous equation of the second order with respect to u is called *Jacobi's equation*.

If a solution of this equation $u = \partial y(x, C)/\partial C$ vanishes at the centre of the pencil, when $x = x_0$ (the centre of the pencil always belongs to the C-discriminant locus), and vanishes at some other point of the interval $x_0 \leqslant x \leqslant x_1$ then the point that is conjugate to A is determined by the equations

$$y = y(x, C_0), \quad \text{and} \quad \frac{\partial y(x, C)}{\partial C} = 0 \quad \text{or} \quad u = 0,$$

and it lies on the extremal arc AB[1].

If there is a solution of Jacobi's equation which vanishes at $x = x_0$ and does not vanish for any other point of the interval $x_0 \leqslant x \leqslant x_1$, then there is no point conjugate to A on the arc AB, i.e. the Jacobi condition is satisfied and the arc AB of an extremal curve can be admitted into a field of extremals with the centre at A.

Remark. It can be proved that the Jacobi condition is a necessary condition for an extremum, i.e. for a curve AB giving an extremum, no conjugate point to the point A can lie in the interval $x_0 < x < x_1$.

EXAMPLE 2. *Is the Jacobi condition satisfied by the extremal curve of the functional*

$$v = \int\limits_0^a (y'^2 - y^2)\, dx$$

which passes through the points $A(0, 0)$ *and* $B(a, 0)$?
The Jacobi equation is

$$-2u - \frac{d}{dx}(2u') = 0 \quad \text{or} \quad u'' + u = 0,$$

and hence

$$u = C_1 \sin(x - C_2).$$

Since $u(0) = 0$, it follows that $C_2 = 0$, and $u = C_1 \sin x$. The function u vanishes at the points $x = k\pi$, where k runs through all integers. Therefore, if $0 < a < \pi$, there is only one point $x = 0$ in the interval $0 \leqslant x \leqslant a$ at which the function u vanishes, and so the Jacobi condition holds. If $a \geqslant \pi$, then there is at least one point more, $x = \pi$, in the interval $0 \leqslant x \leqslant a$, at which the function u vanishes, so that the Jacobi condition does not hold (cf. Example 1, p. 106).

---

[1] We recall that all the non-trivial solutions of a second-order linear homogeneous differential equation satisfying the condition $u(x_0) = 0$ differ with each other only by a constant factor, and therefore they all vanish simultaneously.

EXAMPLE 3. *Is the Jacobi condition satisfied by the extremal curve of the functional*

$$v(y(x)) = \int_0^a (y'^2 + y^2 + x^2)\,dx,$$

which passes through the points $A(0, 0)$ *and* $B(a, 0)$?

The Jacobi equation is $u'' - u = 0$. Its general solution can be put in the form $u = C_1 \sinh x + C_2 \cosh x$. It follows from $u(0) = 0$ that $C_2 = 0$, $u = C_1 \sinh x$. The curves of the pencil $u = C_1 \sinh x$ intersect the x-axis only at one point $x = 0$. The Jacobi condition holds regardless of a.

2. The function $E(x, y, p, y')$

Let us suppose that for the simplest problem of finding extrema of a functional

$$v = \int_{x_0}^{x_1} F(x, y, y')\,dx, \quad y(x_0) = y_0, \quad y(x_1) = y_1,$$

the Jacobi condition holds, and consequently the extremal C which passes through the points $A(x_0, y_0)$ and $B(x_1, y_1)$

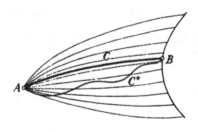

FIG. 42

can be admitted into a central field with a slope function $p(x, y)$ (Fig. 42)[1]. To determine the sign of an increment Δv of a functional v which obtains when passing from an extremal curve C to some other neighbouring admissible

---

[1] It can be assumed that this extremal can be joined not only into a central but also into a proper field.

extremal C^*, we will transform the increment

$$\Delta v = \int\limits_{C^*} F(x, y, y')\, dx - \int\limits_{C} F(x, y, y')\, dx$$

to another form, which will be more convenient to investigate.

The symbols

$$\int\limits_{C^*} F(x, y, y')\, dx \quad \text{and} \quad \int\limits_{C} F(x, y, y')\, dx$$

denote the value of the functional

$$v = \int\limits_{x_0}^{x_1} F(x, y, y')\, dx$$

taken along the arc of C^* or C respectively.

Consider an auxiliary functional

$$\int\limits_{C^*} \left(F(x, y, p) + \left(\frac{dy}{dx} - p \right) F_p(x, y, p) \right) dx$$

which turns into $\int\limits_{C} F(x, y, y')\, dx$ along the extremal C, for $dy/dx = p$ along extremals of the field. On the other hand this same auxiliary functional

$$\int\limits_{C^*} \left(F(x, y, p) + \left(\frac{dy}{dx} - p \right) F_p(x, y, p) \right) dx$$

or

$$(1) \qquad \int\limits_{C^*} \left(F(x, y, p) - p F_p(x, y, p) \right) dx + F_p(x, y, p)\, dy$$

is the integral of an exact differential. In fact, the differential of the function $v^*(x, y)$, which is obtained by restricting the functional $v(y(x))$ to the extremals of the field, has the form (Chapter II, Section 1, p. 69):

$$dv^* = \left(F(x, y, y') - y' F_{y'}(x, y, y') \right) dx + F_{y'}(x, y, y')\, dy$$

and it differs from the integrand of the auxiliary integral (1) only in the symbol used to denote the slope of the tangent to an extremal belonging to the given field.

Consequently the integral

$$\int_{C^*} \left(F+(y'-p)F_p\right)dx$$

coincides along the extremal C with the integral

$$\int_C F(x,y,y')dx,$$

and since the functional

$$\int_{C^*} \left(F+(y'-p)F_p\right)dx$$

is an integral of an exact differential and therefore does not depend on a path of integration, it follows that

$$\int_C F(x,y,y')dx = \int_{C^*} \left(F(x,y,p)+(y'-p)F_p(x,y,p)\right)dx$$

not only when $C^* = C$, but for arbitrary C^*.

Consequently, the increment

$$\Delta v = \int_{C^*} F(x,y,y')dx - \int_C F(x,y,y')dx$$

can be expressed as follows

$$\Delta v = \int_{C^*} F(x,y,y')dx - \int_{C^*} \left(F(x,y,p)+\right.$$
$$\left.+(y'-p)F_p(x,y,p)\right)dx$$
$$= \int_{C^*} \left(F(x,y,y')-F(x,y,p)-(y'-p)F_p(x,y,p)\right)dx.$$

The integrand is called the *Weierstrass function* and it is designated by $E(x,y,p,y')$,

$$E(x,y,p,y')$$
$$= F(x,y,y')-F(x,y,p)-(y'-p)\ F_p(x,y,p).$$

With this notation

$$\Delta v = \int\limits_{x_0}^{x_1} E(x, y, p, y') \, dx.$$

It is obvious that the requirement that the function E be non-negative is a sufficient condition for the functional v to have a minimum along the curve C. For, if $E \geqslant 0$, then $\Delta v \geqslant 0$. A sufficient condition of maximum is that $E \leqslant 0$, for if this inequality holds, then $\Delta v \leqslant 0$. For a weak minimum (cf. p. 21), it is sufficient that the inequality $E(x, y, p, y') \geqslant 0$ (or $E \leqslant 0$ for maximum) holds for all values x, y that are close to the values of x and y along the extremal C, and for all values of y' that are close to $p(x, y)$ along this same extremal. For a strong minimum this same inequality should hold for the same values of x and y, but this time for arbitrary y', for in the case of strong extremum neighbouring curves can have arbitrary directions of tangents, whereas for weak extremum the values of y' along neighbouring curves to the extremal C are close to the value $y' = p$ along the extremal C.

Consequently, the following conditions for a functional v to have an extremum along a curve C are sufficient.

Weak extremum.

1. C is an extremal curve satisfying the boundary conditions.

2. This extremal C can be joined in a field of extremals.

(This condition may be replaced by the Jacobi condition.)

3. The function $E(x, y, p, y')$ has a constant sign for all points (x, y) lying sufficiently close to the curve C and arbitrary values of y' that are sufficiently close to $p(x, y)$. If $E \geqslant 0$ there is a minimum, if $E \leqslant 0$, there is a maximum.

Strong extremum.

1. C is an extremal curve satisfying the boundary conditions.

2. This extremal C can be admitted into a field of extremals.

(This condition may be replaced by the Jacobi condition.)

3. The function $E(x, y, p, y')$ has a constant sign for all points (x, y) lying sufficiently close to the curve C and arbitrary values of y'. If $E \geqslant 0$, there is a minimum. If $E \leqslant 0$, there is a maximum.

Remark. It can be shown that the Weierstrass condition is necessary. More precisely, if for each point lying on an extremal C which belongs to a central field of extremals there are such values of y' for which the function E has opposite signs, then strong extrema do not occur. If this same condition holds with the additional requirement that there are such values of y' that are arbitrarily close to p and for which the function E changes its sign, then weak extrema do not occur either.

EXAMPLE 1. *Examine the extrema of the functional*

$$v = \int_0^a y'^3 dx, \quad y(0) = 0, \quad y(a) = b, \quad a > 0, \quad b > 0.$$

The extremals are straight lines $y = C_1 x + C_2$. An extremum can occur along the line $y = \dfrac{b}{a} x$. The pencil of lines $y = C_1 x$ with centre at the point $(0, 0)$ forms a central field including the extremal $y = \dfrac{b}{a} x$ (Fig. 43).

The function E is

$$E(x, y, p, y') = y'^3 - p^3 - 3p^2(y' - p) = (y' - p)^2(y' + 2p).$$

Along the extremal $y = \dfrac{b}{a} x$ the slope function is $p = b/a > 0$, and if y' takes on such values that are close to $p = b/a$, then $E \geqslant 0$, and consequently the just stated sufficient condition for a weak extremum holds. Therefore there is a weak extremum taken on along the extremal $y = \dfrac{b}{a}x$. If y' may take on arbitrary values, then $y' + 2p$ may have arbitrary sign, and consequently the function E does not keep its sign constant. The sufficient condition for a strong minimum does not hold. Making use of the remark

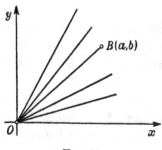

Fig. 43

on p. 115, it can be seen that there is no strong minimum along the line $y = \dfrac{b}{a}x$.

EXAMPLE 2. *Examine the extrema of the functional*

$$\int_0^a (6y'^2 - y'^4 + yy')\,dx, \quad y(0) = 0, \quad y(a) = b, \quad a > 0, \quad b > 0$$

in the class of continuous functions with continuous first derivative.

The extremals are the lines $y = C_1 x + C_2$. The boundary conditions hold of the line $y = \dfrac{b}{a}x$ which can be included in the pencil $y = C_1 x$ of extremals that form a central field. The function E is

$$E(x, y, p, y')$$
$$= 6y'^2 - y'^4 + yy' - 6p^2 + p^4 - yp - (y' - p)(12p - 4p^3 + y)$$
$$= -(y' - p)^2(y'^2 + 2py' - (6 - 3p^2)).$$

Its sign is always opposite to the sign of the last factor

$$(y')^2 + 2py' - (6 - 3p^2).$$

This factor vanishes and may change its sign only when y' passes through the value $y' = -p \pm \sqrt{6 - 2p^2}$. If $6 - 2p^2 \leqslant 0$ or $p \geqslant \sqrt{3}$, then for arbitrary y' we have $y'^2 + 2py' - (6 - 3p^2) \geqslant 0$. If $6 - 2p^2 > 0$ or $p < \sqrt{3}$, then the expression $y'^2 + 2py' - (6 - 3p^2)$ changes its sign. If at the same time y' is very close to p, then the latter expression for $p > 1$ retains the positive sign, and for $p < 1$ it retains the negative sign.

Consequently, if $p = b/a < 1$, or $b < a$, then there is a weak minimum, for $E \geqslant 0$, provided y' is sufficiently close to p. If $p = b/a > 1$ or $b > a$, then there is a weak minimum. If $p = b/a \geqslant \sqrt{3}$ there is a strong maximum, for in this case $E \leqslant 0$ regardless of y'. When $p = b/a < \sqrt{3}$, then it follows from the remark on p. 115 that there is no strong extremum, neither minimum, nor maximum (Fig. 44).

Even in these very simple examples the investigation of the sign of function E was a little tiresome. Therefore we should rather have the condition of constant sign replaced by some other condition that would be easier to

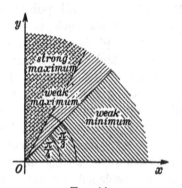

Fig. 44

check. Let us assume that the function $F(x, y, y')$ has third derivative with respect to y'. By Taylor's formula we have

$$F(x, y, y') = F(x, y, p) + (y' - p)F_p(x, y, p) +$$
$$+ \frac{(y' - p)^2}{2!} F_{y'y'}(x, y, q),$$

where q is a value between p and y'.

Replacing $F(x, y, y')$ in the formula

$$E(x, y, p, y') = F(x, y, y') - F(x, y, p) -$$
$$- (y' - p) F_p(x, y, p)$$

by its expression based on Taylor's formula, we obtain

$$E(x, y, p, y') = \frac{(y' - p)^2}{2!} F_{y'y'}(x, y, q).$$

It is easy to see now that the function E has a constant sign, provided $F_{y'y'}(x, y, q)$ has. If we investigate weak extrema, then the function $F_{y'y'}(x, y, q)$ should keep the sign constant for such x, y, q that the points (x, y) lie close to the extremal under investigation and q is close to $p(x, y)$. If $F_{y'y'}(x, y, y') \neq 0$ along the extremal C, then because of the continuity, this second-order derivative keeps the sign constant for all points that are close to the curve C, and for all such values of y' that are close to the values of y' taken at the corresponding points of the curve C. And so, when investigating weak minima the condition $E \geqslant 0$ may be replaced by the inequality $F_{y'y'} > 0$ along a given extremal C. Similarly, when investigating weak maxima the condition $E \leqslant 0$ may be replaced by $F_{y'y'} < 0$ along C. These conditions $F_{y'y'} > 0$, or $F_{y'y'} < 0$ are called the *Legendre conditions*[1].

In the investigation of strong minima the condition $E \geqslant 0$ may be replaced by the requirement that $F_{y'y'}(x, y, q) \geqslant 0$ at the points (x, y) lying close to a given curve under consideration and arbitrary q. It is also necessary to assume then that the Taylor formula

$$F(x, y, y') = F(x, y, p) + (y' - p) F_p(x, y, p) +$$
$$+ \frac{(y' - p)^2}{2!} F_{y'y'}(x, y, q)$$

---

[1] The condition $F_{y'y'} > 0$ or $F_{y'y'} < 0$ is often called the *strong condition of Legendre*, and the *Legendre condition* is then $F_{y'y'} \geqslant 0$ or $F_{y'y'} \leqslant 0$.

holds for arbitrary y'. As a test for strong maxima we obtain the condition $F_{y'y'}(x, y, q) \leqslant 0$, with the same assumptions of the domain of arguments and decomposition of $F(x, y, y')$ by Taylor's formula as before.

EXAMPLE 3. *Examine the extrema of the functional*

$$v(y(x)) = \int_0^a (y'^2 - y^2)\, dx, \quad a > 0, \quad y(0) = 0, \quad y(a) = 0.$$

The Euler equation is $y'' + y = 0$, and its general solution $y = C_1 \cos x + C_2 \sin x$. Making use of boundary conditions we have $C_1 = 0$, $C_2 = 0$, provided $a \neq k\pi$, where k is an arbitrary integer.

Therefore, if $a \neq k\pi$ an extremum may occur only along the line $y = 0$. If $a < \pi$, then the pencil of extremals $y = C_1 \sin x$ with the centre at the point $(0, 0)$ forms a central field. If $a > \pi$ the Jacobi condition does not hold (cf. p. 110-111).

Since the integrand has a third derivative with respect to y' for arbitrary y', and $F_{y'y'} = 2 > 0$ regardless of y', it follows that if $a < \pi$ the straight line $y = 0$ makes v have a strong minimum. Recalling the remark on p. 110, we see that there is no minimum along the line $y = 0$, when $a > \pi$.

EXAMPLE 4. *Examine the extrema of the functional*

$$v(y(x)) = \int_0^{x_1} \frac{\sqrt{1 + y'^2}}{\sqrt{y}}\, dx, \quad y(0) = 0, \quad y(x_1) = y_1.$$

(Cf. the brachistochrone problem on p. 38).

The extremals are cycloids

$$x = C_1(t - \sin t) + C_2, \quad y = C_1^1(1 - \cos t).$$

The pencil of cycloids $x = C_1(t - \sin t)$, $y = C_1(1 - \cos t)$ with centre at the point $(0, 0)$ forms a central field including the extremal

$$x = a(t - \sin t), \quad y = a(1 - \cos t),$$

where a is determined by condition that the desired cycloid should pass through the second boundary point $B(x_1, y_1)$, provided $x_1 < 2\pi a$ (Fig. 45).

We have

$$F_{y'} = \frac{y'}{\sqrt{y}\,\sqrt{1 + y'^2}}, \quad F_{y'y'} = \frac{1}{\sqrt{y}\,(1 + y'^2)^{3/2}} > 0$$

for arbitrary y'. Consequently for $x_1 < 2\pi a$ the cycloid

$$x = a(t - \sin t), \quad y = a(1 - \cos t)$$

gives a strong minimum.

EXAMPLE 5. *Examine the extrema of the functional*

$$v(y(x)) = \int_0^a y'^3 \, dx, \quad y(0) = 0, \quad y(a) = b, \quad a > 0, \quad b > 0.$$

This example was solved on p. 115, but now the investigation for weak extremum can be simplified:

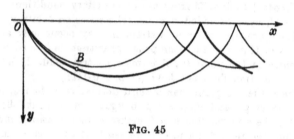

FIG. 45

The extremals are straight lines. The pencil $y = Cx$ forms a central field including $y = \dfrac{b}{a} x$. The second derivative taken along the extremal $y = \dfrac{b}{a} x$ is $F_{y'y'} = 6y' = 6\dfrac{b}{a} > 0$. Consequently the line $y = \dfrac{b}{a} x$ gives a weak minimum. If y' is arbitrary, then the second derivative $F_{y'y'} = 6y'$ does not keep its sign constant. None of the above sufficient conditions for a strong minimum holds. However we cannot conclude that there is no strong extremum at all.

EXAMPLE 6. *Examine the extrema of the functional*

$$v(y(x)) = \int_0^a \frac{y}{y'^2} \, dx, \quad y(0) = 1, \quad y(a) = b, \quad a > 0, \quad 0 < b < 1.$$

The integral curves of the Euler equation (cf. example 5, p. 36) is of the form

$$\frac{y}{y'^2} + y' \frac{2y}{y'^3} = C \quad \text{or} \quad y'^2 = 4C_1 y,$$

taking a square root, separating the variables and then integrating, we obtain $y = (C_1 x + C_2)^2$ which is a family of parabolas. From the condition $y(0) = 1$ we have $C_2 = 1$. The pencil of parabolas $y = (C_1 x + 1)^2$ with centre at the point $A(0, 1)$ has $y = 0$ as its C_1-discriminant locus (Fig. 46). There are two parabolas of this pencil passing through the point $B(a, b)$. There is a conjugate point A^* to the point A, lying on the arc AB of one of these parabolas (L_1). The arc AB of the other parabola (L_2) contains no conjugate point to A. Therefore, the Jacobi condition for the arc L_2 holds and consequently this arc may give an extremum. At a neighbourhood of this extremal arc $F_{y'y'} = 6y/y'^4 > 0$ for

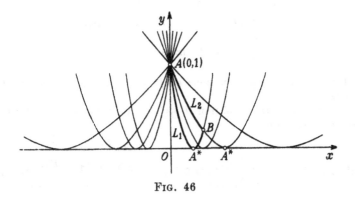

FIG. 46

arbitrary y', yet it does not follow that there is a strong minimum along L_2, for the function $F(x, y, y') = y/y'^2$ cannot be expressed in the form

$$F(x, y, y') = F(x, y, p) + (y' - p) F_p(x, y, p) + $$
$$+ \frac{(y' - p)^2}{2!} F_{y'y'}(x, y, q)$$

for arbitrary y', as there is a discontinuity of the function $F(x, y, y')$ for $y' = 0$. One can only show that there is a weak minimum along L_2, for when y' is sufficiently close to the slope function then it is possible to decompose $F(x, y, y')$ by Taylor's formula. To complete the investigation of this functional for extrema it is necessary to consider the function

$$E(x, y, p, y') = \frac{y}{y'^2} - \frac{y}{p^2} + \frac{2y}{p^3}(y' - p) = \frac{y(y' - p)^2(2y' + p)}{y'^2 p^3}.$$

Since the factor $(2y' + p)$ does not keep its sign constant for arbitrary y', it follows from the remark on p. 115 that there is no strong minimum along L_2.

The theory given above can be extended without any essential changes to functionals of the form

$$v(y_1, y_2, \ldots, y_n) = \int_{x_0}^{x_1} F(x, y_1, y_2, \ldots, y_n, y_1', y_2', \ldots, y_n') \, dx,$$

$$y_i(x_0) = y_{i0}, \quad y_i(x_1) = y_{i1}, \quad i = 1, 2, \ldots, n.$$

The function E takes the form

$$E = F(x, y_1, y_2, \ldots, y_n, y_1', y_2', \ldots, y_n') -$$
$$- F(x, y_1, y_2, \ldots, y_n, p_1, p_2, \ldots, p_n) -$$
$$- \sum_{i=1}^{n} (y_i - p_i) F_{p_i}(x, y_1, y_2, \ldots, y_n, p_1, p_2, \ldots, p_n),$$

where p_i's are slope functions subject to some restricting conditions.

The Legendre condition $F_{y'y'} \geqslant 0$ must be replaced by the set of inequalities

$$F_{y_1'y_1'} \geqslant 0, \quad \begin{vmatrix} F_{y_1'y_1'} & F_{y_1'y_2'} \\ F_{y_2'y_1'} & F_{y_2'y_2'} \end{vmatrix} \geqslant 0, \ldots, \begin{vmatrix} F_{y_1'y_1'} & F_{y_1'y_2'} \cdots F_{y_1'y_n'} \\ F_{y_2'y_1'} & F_{y_2'y_2'} \cdots F_{y_2'y_n'} \\ \cdots \cdots \cdots \cdots \cdots \\ F_{y_n'y_1'} & F_{y_n'y_2'} \cdots F_{y_n'y_n'} \end{vmatrix} \geqslant 0.$$

Sufficient conditions for a weak extremum in simplest problems as well as in more complicated ones, can be also obtained by another way based on investigation of the sign of second variation.

Let us transform by Taylor's formula the increment of a functional involved in a simplest problem as follows

$$\Delta v = \int_{x_0}^{x_1} (F(x, y + \delta y, y' + \delta y') - F(x, y, y')) \, dx$$

$$= \int_{x_0}^{x_1} (F_y \, \delta y + F_{y'} \, \delta y') \, dx +$$

$$+ \frac{1}{2} \int_{x_0}^{x_1} (F_{yy} \, \delta y^2 + 2 F_{yy'} \, \delta y \, \delta y' + F_{y'y'} \, \delta y'^2) \, dx + R,$$

where R is an infinitesimal of order greater than two with respect to δy and $\delta y'$. When we investigate weak extrema, δy and $\delta y'$

are sufficiently small, and so the sign of the increment Δv is determined by the sign of this term of the right member of the equation that involves the lowest powers of δy and $\delta y'$. The first variation vanishes along an extremal,

$$\int_{x_0}^{x_1} (F_y\,\delta y + F_{y'}\,\delta y')\,dx = 0,$$

and consequently the sign of the increment Δv, coincides, in general, with that of second variation

$$\delta^2 v = \int_{x_0}^{x_1} (F_{yy}\,\delta y^2 + 2F_{yy'}\,\delta y\,\delta y' + F_{y'y'}\,\delta y'^2)\,dx.$$

The Legendre condition along with Jacobi's condition guarantee that the sign of second variation is constant, and consequently they guarantee that the sign of the increment Δv is constant in a problem of finding weak extrema.

In fact, let us consider the integral

$$(*) \qquad \int_{x_0}^{x_1} (\omega'(x)\,\delta y^2 + 2\omega(x)\,\delta y\,\delta y')\,dx,$$

where $\omega(x)$ is an arbitrary differentiable function. This integral vanishes

$$\int_{x_0}^{x_1} (\omega'(x)\,\delta y^2 + 2\omega(x)\,\delta y\,\delta y')\,dx = \int_{x_0}^{x_1} \frac{d}{dx}(\omega\,\delta y^2)\,dx = [\omega(x)\,\delta y^2]_{x_0}^{x_1} = 0,$$

for $\delta y|_{x_0} = \delta y|_{x_1} = 0$.

Adding the integral $(*)$ to the second variation, we find that

$$\delta^2 v = \int_{x_0}^{x_1} ((F_{yy} + \omega')\,\delta y^2 + 2(F_{yy'} + \omega)\,\delta y\,\delta y' + F_{y'y'}\,\delta y'^2)\,dx.$$

We shall choose the function $\omega(x)$ so that the integrand will turn into a perfect square, up to a certain factor. To this aim the function $\omega(x)$ should satisfy the equation

$$F_{y'y'}(F_{yy} + \omega') - (F_{yy'} + \omega)^2 = 0.$$

If ω is so chosen, the second variation takes the form

$$\delta^2 v = \int_{x_0}^{x_1} F_{y'y'}\left(\delta y' + \frac{F_{yy'} + \omega}{F_{y'y'}}\,\delta y\right)^2 dx,$$

and consequently the sign of the second variation and the sign of $F_{y'y'}$ are the same.

The list of sufficient conditions

$$v(y(x)) = \int_{x_0}^{x_1} F(x, y, y')\, dx,$$

Weak minimum	Strong minimum	Weak minimum
1. $F_y - \dfrac{d}{dx} F_{y'} = 0$	1. $F_y - \dfrac{d}{dx} F_{y'} = 0$	1. $F_y - \dfrac{d}{dx} F_{y'} = 0$
2. Jacobi condition	2. Jacobi condition	2. Jacobi condition
3. $F_{y'y'} > 0$ along the investigated extremal.	3. $F_{y'y'}(x, y, y') \geqslant 0$ for those points (x,y) that are close to the extremal under examination and for arbitrary values of y'. It is assumed here that $F(x, y, y')$ has third-order derivative with respect to y', for all y'.	3. $E(x, y, p, y') \geqslant 0$ for all the points (x, y) sufficiently close to the extremal under examination, and all y' sufficiently close to $p(x, y)$.

(^{1}) To obtain sufficient conditions for a maximum we have only to change the sense of inequalities.

However such a transformation is permitted only if the differential equation

$$F_{y'y'}(\omega' + F_{yy}) - (F_{yy'} + \omega)^2 = 0$$

has a differentiable solution $\omega(x)$ in the interval (x_0, x_1).

Performing a transformation to a new set of variables substituting

$$\omega = -F_{yy'} - F_{y'y'}\frac{u'}{u},$$

for a minimum of a simplest functional[1]

$$y(x_0) = y_0, \quad y(x_1) = y_1$$

Strong minimum	Weak minimum	Strong minimum
1. $F_y - \dfrac{d}{dx} F_{y'} = 0$	1. $F_y - \dfrac{d}{dx} F_{y'} = 0$	1. $F_y - \dfrac{d}{dx} F_{y'} = 0$
2. Jacobi condition	2. There exists a field of extremals including the investigated extremal curve.	2. There exists a fields of extremals including the investigated extremal curve.
3. $E(x, y, p, y') \geqslant 0$ for all the points (x, y) sufficiently close to the extremal under examination, and arbitrary y'.	3. $E(x, y, p\ y') \geqslant 0$ for all the points (x, y) sufficiently close to the extremal under examination and all y' sufficiently close to $p(x, y)$.	3. $E(x, y, p, y') \geqslant 0$ for all the points (x, y) sufficiently close to the extremal under examination, and arbitrary y'.

where u is a new unknown function, we obtain

$$\left(F_{yy} - \frac{d}{dx} F_{yy'}\right) u - \frac{d}{dx}(F_{y'y'} u') = 0$$

and this is exactly the Jacobi equation (cf. p. 108).

If a solution of this equation which is non-vanishing on the interval $x_0 < x \leqslant x_1$ exists, i. e. if Jacobi condition holds, then there exists for the same values of x a continuous differentiable solution $\omega(x) = -F_{yy'} - F_{y'y'} u'/u$ of the equation

$$F_{y'y'}(F_{yy} + \omega') - (F_{yy'} + \omega)^2 = 0.$$

And so the Legendre and Jacobi conditions ensure that the sign of the second variation is constant.

Problems

Examine the extrema of the functionals:

1. $v(y(x)) = \int_0^2 (xy' + y'^2)\,dx$, $y(0) = 1$, $y(2) = 0$.

2. $v(y(x)) = \int_0^a (y'^2 + 2yy' - 16y^2)\,dx$, $a > 0$, $y(0) = 0$, $y(a) = 0$.

3. $v(y(x)) = \int_{-1}^2 y'(1 + x^2 y')\,dx$, $y(-1) = 1$, $y(2) = 4$.

4. $v(y(x)) = \int_1^2 y'(1 + x^2 y')\,dx$, $y(1) = 3$, $y(2) = 5$.

5. $v(y(x)) = \int_{-1}^2 y'(1 + x^2 y')\,dx$, $y(-1) = y(2) = 1$.

6. $v(y(x)) = \int_0^{\pi/4} (4y^2 - y'^2 + 8y)\,dx$, $y(0) = -1$, $y(\pi/4) = 0$.

7. $v(y(x)) = \int_1^2 (x^2 y'^2 + 12y^2)\,dx$, $y(1) = 1$, $y(2) = 8$.

8. $v(y(x)) = \int_{x_0}^{x_1} \frac{1 + y^2}{y'^2}\,dx$, $y(x_0) = y_0$, $y(x_1) = y_1$.

9. $v(y(x)) = \int_0^1 (y'^2 + y^2 + 2ye^{2x})\,dx$, $y(0) = \frac{1}{3}$, $y(1) = \frac{1}{3}e^2$.

10. $v(y(x)) = \int_0^{\pi/4} (y^2 - y'^2 + 6y\sin 2x)\,dx$, $y(0) = 0$, $y(\pi/4) = 1$.

11. $v(y(x)) = \int_0^{x_1} \frac{dx}{y'}$, $y(0) = 0$, $y(x_1) = y_1$, $x_1 > 0$, $y_1 > 0$.

12. $vy((x)) = \int_0^{x_1} \frac{dx}{y'^2}$, $y(0) = 0$, $y(x_1) = y_1$, $x_1 > 0$, $y_1 > 0$.

13. $v(y(x)) = \int_1^2 \frac{x^3}{y'^2}\,dx$, $y(1) = 1$, $y(2) = 4$.

14. $v(y(x)) = \int_1^3 (12xy + y'^2)\,dx$, $y(1) = 0$, $y(3) = 26$.

CHAPTER IV

VARIATIONAL PROBLEMS
OF CONSTRAINED EXTREMA

1. Constraints of the form $\varphi(x, y_1, y_2, \ldots, y_n) = 0$

We shall now consider the problems that consist in finding extrema of a functional v, where the functional arguments of v are subject to some additional constraining relations. Such extrema are called *constrained extrema*. For instance, one such problem is to examine the extrema of a functional

$$v(y_1, y_2, \ldots, y_n) = \int_{x_0}^{x_1} F(x, y_1, y_2, \ldots, y_n, y_1', y_2', \ldots, y_n')\, dx$$

with the additional condition that

$$\varphi_i(x, y_1, y_2, \ldots, y_n) = 0, \quad i = 1, 2, \ldots, m, \quad m < n.$$

Let us recall the method of solution of an analogous problem of finding the extrema of an ordinary function $z = f(x_1, x_2, \ldots, x_n)$, with constraining relations

$$\varphi_i(x_1, x_2, \ldots, x_n) = 0, \quad i = 1, 2, \ldots, m, \quad m < n.$$

The original method consists first in solving the system of equations

$$\varphi_i(x_1, x_2, \ldots, x_n) = 0, \quad i = 1, 2, \ldots, m$$

which we will consider independent, with respect to any m out of n arguments x_i, for instance, first m arguments

$x_1, x_2, \ldots, x_m$:

$$x_1 = x_1(x_{m+1}, x_{m+2}, \ldots, x_n),$$
$$x_2 = x_2(x_{m+1}, x_{m+2}, \ldots, x_n),$$
$$\cdot \cdot \cdot \cdot \cdot \cdot \cdot \cdot \cdot \cdot \cdot \cdot \cdot \cdot,$$
$$x_m = x_m(x_{m+1}, x_{m+2}, \ldots, x_n),$$

and then substituting the expressions so obtained for $x_1, x_2, \ldots, x_m$ into the expression $f(x_1, x_2, \ldots, x_n)$. By so doing this latter function turns into a function $\Phi(x_{m+1}, x_{m+2}, \ldots, x_n)$ of $n-m$ variables $x_{m+1}, x_{m+2}, \ldots, x_n$, which are now independent. Therefore the original problem has been reduced to that of finding ordinary extrema of the function Φ. In the same way the variational problem formulated above can be solved. By solving the system $\varphi_i(x, y_1, y_2, \ldots, y_n) = 0$, $i = 1, 2, \ldots, m$, with respect to $y_1, y_2, \ldots, y_m$ or with respect to any other set of m functions from among y_i, and substituting their expressions into $v(y_1, y_2, \ldots, y_n)$, a functional $W(y_{m+1}, y_{m+2}, \ldots, y_n)$ obtains, depending only on $n-m$ independent arguments. Consequently, this functional can be investigated by the methods described in Section 3, Chapter I. However, both for functions and for functionals, it is usually more convenient to follow another method, the *Lagrange method of multipliers*, which does not make any distinction among the variables. As is well known, if $z = f(x_1, x_2, \ldots, x_n)$ is a function to be examined for extrema with constraining relations $\varphi_i(x_1, x_2, \ldots, x_n) = 0$, $i = 1, 2, \ldots, m$, then Lagrange method of multipliers consists in considering an auxiliary function

$$z^* = f + \sum_{i=1}^{m} \lambda_i \varphi_i,$$

where λ_i are some constant multipliers, and then finding the ordinary extrema of the function z^*. We have, therefore, to solve the system of equations $\partial z^*/\partial x_j = 0$, $j = 1, 2, \ldots, n$, along with the equations of constraints

$\varphi_i = 0$, $i = 1, 2, \ldots, m$, to determine all $n+m$ unknown quantities $x_1, x_2, \ldots, x_n$, and $\lambda_1, \lambda_2, \ldots, \lambda_m$. The problems of constrained extrema of functionals can be solved likewise.

If

$$v = \int\limits_{x0}^{x_1} F(x, y_1, y_2, \ldots, y_n, y'_1, y'_2, \ldots, y'_n)\, dx$$

and

$$\varphi_i(x, y_1, y_2, \ldots, y_n) = 0, \quad i = 1, 2, \ldots, m,$$

then we have to set the functional

$$v^* = \int\limits_{x_0}^{x_1} \Big(F + \sum_{i=1}^{m} \lambda_i(x)\varphi_i\Big) dx$$

or

$$v^* = \int\limits_{x_0}^{x_1} F^*\, dx,$$

where

$$F^* = F + \sum_{i=1}^{m} \lambda_i(x)\varphi_i,$$

and to examine it for ordinary extrema, i. e. solve the system of Euler equations

$$\text{(I)} \qquad F^*_{y_j} - \frac{d}{dx} F^*_{y'_j} = 0, \quad j = 1, 2, \ldots, n,$$

with additional equations of constraints $\varphi_i = 0$, $i = 1, 2, \ldots, m$.

In general these $m+n$ equations are enough to determine $m+n$ unknown functions $y_1, y_2, \ldots, y_n$ and $\lambda_1, \lambda_2, \ldots, \lambda_m$, and the boundary conditions $y_j(x_0) = y_{j0}$ and $y_j(x_1) = y_{j1}$, $j = 1, 2, \ldots, n$, which shall be compatible with the constraining relations, can be used to determine $2n$

arbitrary constants in the general solution of the system of Euler equations.

It is obvious that the curves so obtained that make the functional v^* have minimum or maximum, are the solution of the original variational problem, for if the functions

$$\lambda_i(x), \quad i = 1, 2, \ldots, m, \quad \text{and} \quad y_j, \, j = 1, 2, \ldots, n,$$

satisfy the system of equations (I), then

$$\varphi_i = 0 \quad \text{for all} \quad i = 1, 2, \ldots, m,$$

and consequently $v^* = v$. Now if there is an ordinary extremum of v^* along the curve $y_j = y_j(x), j = 1, 2, \ldots, n$, which is determined from the system (I), then it is an extremum in the class of all neighbouring curves that may satisfy the equations of constraints or not, and consequently it is also an extremum in a narrower class of neighbouring curves that satisfy the equations of constraints.

It does not follow from this, however, that all solutions of the original problem of constrained extrema make the functional v^* have an ordinary extremum, and so it is not clear if all solutions of the original problem can be obtained this way. We shall prove only a weaker result.

THEOREM. *A sequence of functions $y_1, y_2, \ldots, y_n$ which make the functional*

$$v = \int_{x_0}^{x_1} F(x, y_1, y_2, \ldots, y_n, y_1', y_2', \ldots, y_n') \, dx$$

have an extremum, with constraints

$$\varphi_i(x, y_1, y_2, \ldots, y_n) = 0, \quad i = 1, 2, \ldots, m, \, m < n,$$

satisfies for suitably chosen multipliers $\lambda_i(x), i = 1, 2, \ldots, m$, the Euler equations for the functional

$$v^* = \int_{x_0}^{x_1} \left(F + \sum_{i=1}^{m} \lambda_i(x) \varphi_i \right) dx = \int_{x_0}^{x_1} F^* \, dx.$$

The functions $\lambda_i(x)$ and $y_i(x)$ can be determined from the Euler equations

$$F_{y_j}^* - \frac{d}{dx} F_{y_j'}^* = 0, \quad j = 1, 2, \ldots, n,$$

and the equations of constraints

$$\varphi_i = 0, \quad i = 1, 2, \ldots, m.$$

The equations $\varphi_i = 0$ can also be considered Euler equations for the functional v^*, if we consider not only $y_1, y_2, \ldots, y_n$ as arguments of v^*, but also $\lambda_1(x), \lambda_2(x), \ldots, \lambda_m(x)$. The equations

$$\varphi_i(x, y_1, y_2, \ldots, y_n) = 0, \quad i = 1, 2, \ldots, m,$$

are assumed to be independent, i.e. there is a non-vanishing Jacobian of order m, for instance

$$\frac{D(\varphi_1, \varphi_2, \ldots, \varphi_m)}{D(y_1, y_2, \ldots, y_m)} \neq 0.$$

Proof. The fundamental condition of an extremum $\delta v = 0$ takes the form

$$\int_{x_0}^{x_1} \sum_{j=1}^{n} (F_{y_j} \delta y_j + F_{y_j'} \delta y_j') \, dx = 0$$

or, integrating the second term in each bracket by parts, and bearing in mind that

$$(\delta y_j)' = \delta y_j' \quad \text{and} \quad (\delta y_j)_{x=x_0} = 0, \quad (\delta y_j)_{x=x_1} = 0,$$

we obtain

$$\int_{x_0}^{x_1} \sum_{j=1}^{n} \left(F_{y_j} - \frac{d}{dx} F_{y_j'} \right) \delta y_j \, dx = 0.$$

Since $y_1, y_2, \ldots, y_n$ are subject to m independent constraining relations

$$\varphi_i(x, y_1, y_2, \ldots, y_n) = 0, \quad i = 1, 2, \ldots, m,$$

the variations δy_j are not arbitrary, and consequently the fundamental lemma cannot be applied. The variations δy_j must satisfy the following relations, obtained by varying the constraining relations $\varphi_i = 0$:

$$\sum_{j=1}^{n} \frac{\partial \varphi_i}{\partial y_j} \, \delta y_j = 0, \quad i = 1, 2, \ldots, m \; (^1),$$

and consequently only $n - m$ from among the variations δy_j can be considered independent, for instance $\delta y_{m+1}, \delta y_{m+2}, \ldots, \delta y_n$ and the remaining ones are to be determined from the equations just obtained.

Multiplying successive equations by $\lambda_i(x) \, dx$, with corresponding i, and integrating from x_0 to x_1, we obtain

$$\int_{x_0}^{x_1} \lambda_i(x) \sum_{j=1}^{n} \frac{\partial \varphi_i}{\partial y_j} \delta y_j \, dx = 0, \quad i = 1, 2, \ldots, m.$$

Adding all these m equations which are satisfied by all the admitted variations δy_j, along with the equation

$$\int_{x_0}^{x_1} \sum_{j=1}^{n} \left(F_{y_j} - \frac{d}{dx} F_{y'_j} \right) \delta y_j \, dx = 0,$$

(¹) To be more precise, applying the Taylor formula to the difference

$$\varphi_i(x, y_1 + \delta y_1, \ldots, y_n + \delta y_n) - \varphi_i(x, y_1, \ldots, y_n)$$

of the left-hand numbers of the equations $\varphi_i(x, y_1 + \delta y_1, \ldots, y_n + \delta y_n) = 0$ and $\varphi_i(x, y_1, y_2, \ldots, y_n) = 0$, we should have written

$$\sum_{j=1}^{n} \frac{\partial \varphi_i}{\partial y_j} \delta y_j + R_i = 0,$$

where R_i has order higher than one with respect to δy_j, $j = 1, 2, \ldots, n$. However, it is easy to check that the terms R_i do not play any rôle in the sequel, for calculating the variation of a functional we are interested only in the first-order terms with respect to δy_j, $j = 1, 2, \ldots, n$.

we have

$$\int_{x_0}^{x_1} \sum_{j=1}^{n} \left(\frac{\partial F}{\partial y_j} + \sum_{i=1}^{m} \lambda_i(x) \frac{\partial \varphi_i}{\partial y_j} - \frac{d}{dx} \frac{\partial F}{\partial y_j'} \right) \delta y_j \, dx = 0.$$

Setting

$$F^* = F + \sum_{i=1}^{m} \lambda_i(x) \varphi_i$$

we then have

$$\int_{x_0}^{x_1} \sum_{j=1}^{n} \left(F_{y_j}^* - \frac{d}{dx} F_{y_j'}^* \right) \delta y_j \, dx = 0.$$

Again, since the variations δy_j are not independent, we cannot apply the fundamental lemma. We choose m multipliers $\lambda_1(x), \lambda_2(x), \ldots, \lambda_m(x)$ in such a way that they satisfy m equations

$$F_{y_j}^* - \frac{d}{dx} F_{y_j'}^* = 0, \quad j = 1, 2, \ldots, m,$$

or

$$\frac{\partial F}{\partial y_j} + \sum_{i=1}^{m} \lambda_i(x) \frac{\partial \varphi_i}{\partial y_j} - \frac{d}{dx} \frac{\partial F}{\partial y_j'} = 0, \quad j = 1, 2, \ldots, m.$$

This is a system of linear equations with respect to λ_i with non-vanishing Jacobian

$$\frac{D(\varphi_1, \varphi_2, \ldots, \varphi_m)}{D(y_1, y_2, \ldots, y_m)} \neq 0.$$

Consequently, there is a solution of this system

$$\lambda_1(x), \lambda_2(x), \ldots, \lambda_m(x).$$

With this choice of $\lambda_1(x), \lambda_2(x), \ldots, \lambda_m(x)$ the fundamental necessary condition for an extremum

$$\int_{x_0}^{x_1} \sum_{j=1}^{n} \left(F_{y_j}^* - \frac{d}{dx} F_{y_j'}^* \right) \delta y_j \, dx = 0$$

takes the form

$$\int_{x_0}^{x_1} \sum_{j=m+1}^{n} \left(F_{y_j}^* - \frac{d}{dx} F_{y'_j}^* \right) \delta y_j \, dx = 0$$

and since for functions $y_1, y_2, \ldots, y_n$ that make the functional v have an extremum this functional equation turns into identity even if δy_j, $j = m+1, m+2, \ldots, n$, are arbitrary, it follows that the fundamental lemma can be applied. Setting all δy_j equal to zero, save one of them, and applying the lemma, and repeating this procedure $n-m$ times we have

$$F_{y_j}^* - \frac{d}{dx} F_{y'_j}^* = 0, \quad j = m+1, m+2, \ldots, n.$$

Remembering that

$$F_{y_j}^* - \frac{d}{dx} F_{y'_j}^* = 0, \quad j = 1, 2, \ldots, m,$$

we finally come to the conclusion that the functions which make the functional v have a constrained extremum, along with the multipliers $\lambda_i(x)$ shall satisfy the system of equations

$$F_{y_j}^* - \frac{d}{dx} F_{y'_j}^* = 0, \quad j = 1, 2, \ldots, n,$$

$$\varphi_i(x, y_1, y_2, \ldots, y_n) = 0, \quad i = 1, 2, \ldots, m.$$

EXAMPLE 1. *Find the shortest distance between two points $A(x_0, y_0, z_0)$ and $B(x_1, y_1, z_1)$ lying on a surface $\varphi(x, y, z) = 0$.* (Cf. the problem of geodesics, p. 11.) As it is well known, the distance between two points on a surface is given by the formula

$$l = \int_{x_0}^{x_1} \sqrt{1 + y'^2 + z'^2} \, dx.$$

In the present case we have to find a minimum of l with constraining relation $\varphi(x, y, z) = 0$. As has been just indicated, we consider the auxiliary functional

$$l^* = \int\limits_{x_0}^{x_1} \left(\sqrt{1 + y'^2 + z'^2} + \lambda(x) \varphi(x, y, z) \right) dx$$

and write down its Euler equations:

$$\lambda(x)\varphi_y - \frac{d}{dx} \frac{y'}{\sqrt{1 + y'^2 + z'^2}} = 0,$$

$$\lambda(x)\varphi_z - \frac{d}{dx} \frac{z'}{\sqrt{1 + y'^2 + z'^2}} = 0, \quad \varphi(x, y, z) = 0.$$

This system of equations determines both the unknown functions

$$y = y(x), \quad z = z(x)$$

that may give a constrained extremum of the functional l, as well as the multiplier $\lambda(x)$.

EXAMPLE 2. *Using the Ostrogradski-Hamilton principle* (cf. p. 58-59) *find the equations of motion for a system of material points with masses* m_i, $i = 1, 2, \ldots, n$, *and with coordinates* (x_i, y_i, z_i) *when the force acting on them has a force function* U, *and the coordinates* (x_i, y_i, z_i) *and the time t are subject to constraining relations*

$$\varphi_j(t, x_1, x_2, \ldots, x_n, y_1, y_2, \ldots, y_n, z_1, z_2, \ldots, z_n) = 0,$$

$$j = 1, 2, \ldots, m.$$

The Ostrogradski-Hamilton integral

$$v = \int\limits_{t_0}^{t_1} (T - U) dt$$

is in this case of the form

$$v = \int\limits_{t_0}^{t_1} \left(\tfrac{1}{2} \sum_{i=1}^{n} m_i (\dot{x}_i^2 + \dot{y}_i^2 + \dot{z}_i^2) - U \right) dt$$

and the auxiliary functional is

$$v^* = \int\limits_{t_0}^{t_1} \left(\tfrac{1}{2} \sum_{i=1}^{n} m_i (\dot{x}_i^2 + \dot{y}_i^2 + \dot{z}_i^2) - U + \sum_{j=1}^{m} \lambda_j(t) \varphi_j \right) dt.$$

The equations of motion are the Euler equations of the functional v^*; they take the following form

$$m_i x_i{}'' = -\frac{\partial U}{\partial x_i} + \sum_{j=1}^{m} \lambda_j(t)\frac{\partial \varphi_j}{\partial x_i},$$

$$m_i y_i{}'' = -\frac{\partial U}{\partial y_i} + \sum_{j=1}^{m} \lambda_j(t)\frac{\partial \varphi_j}{\partial y_i},$$

$$m_i z_i{}'' = -\frac{\partial U}{\partial z_i} + \sum_{j=1}^{m} \lambda_j(t)\frac{\partial \varphi_j}{\partial z_i}, \quad i = 1, 2, \ldots, n.$$

2. Constraints of the form $\varphi(x, y_1, y_2, \ldots, y_n, y_1', y_2', \ldots, y_n') = 0$

In the preceding section we considered the problem of finding extrema of the functional

$$v = \int_{x_0}^{x_1} F(x, y_1, y_2, \ldots, y_n, y_1', y_2', \ldots, y_n')\,dx,$$

$$y_j(x_0) = y_{j0}, \quad y_j(x_1) = y_{j1}, \quad j = 1, 2, \ldots, n,$$

with finite constraints

$$\varphi_i(x, y_1, y_2, \ldots, y_n) = 0, \quad i = 1, 2, \ldots, m.$$

We will now assume that the constraining relations are differential equations

$$\varphi_i(x, y_1, y_2, \ldots, y_n, y_1', y_2', \ldots, y_n') = 0, \quad i = 1, 2, \ldots, m.$$

In this case also a principle of undetermined multipliers can be proved, stating that constrained extrema of the functional v can occur only along those curves that give ordinary extrema of the functional v^*

$$v^* = \int_{x_0}^{x_1} \Big(F + \sum_{i=1}^{m} \lambda_i(x)\varphi_i\Big)\,dx = \int_{x_0}^{x_1} F^*\,dx,$$

where

$$F^* = F + \sum_{i=1}^{m} \lambda_i(x)\varphi_i.$$

However, the proof is much more complicated than that for finite constraints.

If we wish to prove a weaker assertion, that the curves along which the functional v has constrained extrema are extremals of the functional v^*, for suitably chosen $\lambda_i(x)$, then we can follow the proof of the analogous assertion in the preceding section with only slight changes.

In fact, let us assume that there is a non-vanishing functional determinant of order m, for instance

$$\frac{D(\varphi_1, \varphi_2, \ldots, \varphi_m)}{D(y_1', y_2', \ldots, y_m')} \neq 0.$$

This means that the constraints are independent.

Solving the equations $\varphi_i(x, y_1, y_2, \ldots, y_n, y_1', y_2', \ldots, y_n') = 0$ with respect to $y_1', y_2', \ldots, y_n'$, which is of course possible, for

$$\frac{D(\varphi_1, \varphi_2, \ldots, \varphi_m)}{D(y_1', y_2', \ldots, y_m')} \neq 0,$$

we obtain

$$y_i' = \psi_i(x, y_1, \ldots, y_n, y_{m+1}', \ldots, y_n'),$$

$i = 1, 2, \ldots, m$. If we consider the functions $y_{m+1}, y_{m+2}, \ldots, y_n$ to be independent, then this system of differential equations determines $y_1, y_2, \ldots, y_m$. The functions $y_{m+1}, y_{m+2}, \ldots, y_n$ are arbitrary differentiable functions with fixed boundary conditions, and therefore their variations are, with this meaning, arbitrary variations.

Let $y_1, y_2, \ldots, y_n$ be an arbitrary system of admissible functions satisfying the constraining relations $\varphi_i = 0$, $i = 1, 2, \ldots, m$.

Let us vary the equations of constraints

$$\sum_{j=1}^{n} \frac{\partial \varphi_i}{\partial y_j} \delta y_j + \sum_{j=1}^{n} \frac{\partial \varphi_i}{\partial y_j'} \delta y_j' = 0, \quad i = 1, 2, \ldots, m \ (^1).$$

We multiply successively each of these equations by an as yet arbitrary multiplier $\lambda_i(x)$ and then integrate from x_0 to x_1. We

---

(^{1}) As on p. 132 one should add up to the left-hand members of these equations some expressions that consist of terms of order greater than one with respect to δy_j and $\delta y_j'$, $j = 1, 2, \ldots, n$, but in the present case it is much more difficult to estimate the influence of these non-linear terms.

obtain the equation

$$\int\limits_{x_0}^{x_1} \lambda_i(x) \sum_{j=1}^{n} \frac{\partial \varphi_i}{\partial y_j}\, \delta y_j\, dx + \int\limits_{x_0}^{x_1} \lambda_i(x) \sum_{j=1}^{n} \frac{\partial \varphi_i}{\partial y_j'}\, \delta y_j'\, dx = 0.$$

Integrating each term of the first integral by parts and bearing in mind that $\delta y_j' = (\delta y_j)'$ and $(\delta y_j)_{x=x_0} = (\delta y_j)_{x=x_1} = 0$, we have

(I) $$\int\limits_{x_0}^{x_1} \sum_{j=1}^{n} \left(\lambda_i(x) \frac{\partial \varphi_i}{\partial y_j} - \frac{d}{dx}\left(\lambda_i(x) \frac{\partial \varphi_i}{\partial y_j'} \right) \right) \delta y_j\, dx = 0.$$

It follows from the fundamental necessary condition of an extremum, $\delta v = 0$, that

(II) $$\int\limits_{x_0}^{x_1} \sum_{j=1}^{n} \left(F_{y_j} - \frac{d}{dx} F_{y_j'} \right) \delta y_j\, dx = 0$$

for

$$\delta v = \int\limits_{x_0}^{x_1} \sum_{j=1}^{n} (F_{y_j} \delta y_j + F_{y_j'} \delta y_j')\, dx = \int\limits_{x_0}^{x_1} \sum_{j=1}^{n} \left(F_{y_j} - \frac{d}{dx} F_{y_j'} \right) \delta y_j\, dx.$$

Adding all the equations (I) and the equation (II) and setting $F^* = F + \sum\limits_{i=1}^{m} \lambda_i(x)\varphi_i$ we have

(III) $$\int\limits_{x_0}^{x_1} \sum_{j=1}^{n} \left(F_{y_j}^* - \frac{d}{dx} F_{y_j'}^* \right) \delta y_j\, dx = 0.$$

Since the variations δy_j, $j = 1, 2, \ldots, n$, are not independent, we cannot apply the fundamental lemma. We will now choose m multipliers $\lambda_1(x), \lambda_2(x), \ldots, \lambda_m(x)$ in such a way that

$$F_{y_j}^* - \frac{d}{dx} F_{y_j'}^* = 0, \quad j = 1, 2, \ldots, m.$$

If we write out all these equations in full, then it is plain that they are a system of linear differential equations with respect to

$$\lambda_i(x) \quad \text{and} \quad \frac{d\lambda_i}{dx}, \quad i = 1, 2, \ldots, m,$$

which, under the assumptions already made, has a solution $\lambda_1(x)$, $\lambda_2(x), \ldots, \lambda_m(x)$ depending on m arbitrary constants. Substituting these $\lambda_1(x), \lambda_2(x), \ldots, \lambda_m(x)$ into the equation (III), we obtain

$$\int\limits_{x_0}^{x_1} \sum_{j=m+1}^{n} \left(F_{y_j}^* - \frac{d}{dx} F_{y_j'}^* \right) \delta y_j \, dx = 0,$$

where the variations δy_j, $j = m+1, m+2, \ldots, n$ are arbitrary, and consequently, setting all variations $\delta y_j = 0$ but one of them, say δy_j, which is left arbitrary and applying the fundamental lemma we obtain

$$F_{y_j}^* - \frac{d}{dx} F_{y_j'}^* = 0, \quad j = m+1, m+2, \ldots, n.$$

Consequently the functions $y_1(x), y_2(x), \ldots, y_n(x)$ which make the functional v take a constrained extremum, as well as the multipliers $\lambda_1(x), \lambda_2(x), \ldots, \lambda_m(x)$ must satisfy the system of $n+m$ equations

$$F_{y_j}^* - \frac{d}{dx} F_{y_j'}^* = 0, \quad j = 1, 2, \ldots, n,$$

and

$$\varphi_i = 0, \quad i = 1, 2, \ldots, m,$$

i.e. they must satisfy the Euler equations of the auxiliary functional v^*, which shall be considered dependent on $n+m$ functions $y_1, y_2, \ldots, y_n, \lambda_1, \lambda_2, \ldots, \lambda_m$.

3. Isoperimetric problems

In the narrowest sense, the *isoperimetric problem* is the problem of finding the geometric figure with maximal area and given perimeter.

This extremal problem which was studied in ancient Greece is, of course, a variational problem, as was mentioned above on p. 12 ([1]).

Representing the curve in parametric form $x = x(t)$, $y = y(t)$, we can formulate this problem as follows: find

_____

([1]) Though the solution of this problem was known to the ancient Greeks, its variational character was first described at the end of the XVIIth century.

the maximum of the functional

$$S = \int_{t_0}^{t_1} xy^{\cdot} dt \quad (\text{or } S = \tfrac{1}{2} \int_{t_0}^{t_1} (xy^{\cdot} - yx^{\cdot}) dt)$$

with the additional condition that the functional

$$\int_{t_0}^{t_1} \sqrt{x^{\cdot 2} + y^{\cdot 2}} \, dt$$

has the constant value l. Consequently, we have a variational problem for a constrained extremum subject to a peculiar constraining relation, namely, that the integral $\int_{t_0}^{t_1} \sqrt{x^{\cdot 2} + y^{\cdot 2}} \, dt$ is constant.

Nowadays a much wider class of problems is called *isoperimetric problems*, namely, all such problems where one has to find extrema of a functional

$$v = \int_{x_0}^{x_1} F(x, y_1, y_2, \ldots, y_n, y_1', y_2', \ldots, y_n') dx$$

while keeping some other functionals constant, for instance

$$\int_{x_0}^{x_1} F_i(x, y_1, y_2, \ldots, y_n, y_1', y_2', \ldots, y_n') dx = l_i,$$

$$i = 1, 2, \ldots, m,$$

where l_i are constants, m being greater, less, or equal to n, and analogous problems for more complicated functionals. The constraining relations of this kind are called *isoperimetric conditions*.

By introducing new unknown functions, isoperimetric problems can be reduced to problems of constrained extrema, considered in the preceding section.

Let us define functions $z_i(x)$ as follows

$$z_i(x) = \int_{x_0}^{x} F_i \, dx, \quad i = 1, 2, \ldots, m,$$

so that $z_i(x_0) = 0$, and $z_i(x_1) = l_i$, for $\int_{x_0}^{x_1} F_i \, dx = l_i$.

Differentiating z_i with respect to x, we have

$$z_i'(x) = F_i(x, y_1, y_2, \ldots, y_n, y_1', y_2', \ldots, y_n'),$$
$$i = 1, 2, \ldots, m.$$

In this way the integral isoperimetric constraints have been replaced by the differential constraints

$$z_i' = F_i(x, y_1, y_2, \ldots, y_n, y_1', y_2', \ldots, y_n'), \quad i = 1, 2, \ldots, m,$$

and the problem has been reduced in this way to that considered in the preceding section.

Applying the rule of Lagrangean multipliers, instead of investigating the constrained extrema of the functional

$$v = \int_{x_0}^{x_1} F \, dx$$

with the constraints $F_i - z_i' = 0$, $i = 1, 2, \ldots, m$, we can investigate the ordinary extrema of the functional

$$v^* = \int_{x_0}^{x_1} \left(F + \sum_{i=1}^{m} \lambda_i(x)(F_i - z_i') \right) dx = \int_{x_0}^{x_1} F^* \, dx,$$

where

$$F^* = F + \sum_{i=1}^{m} \lambda_i(x)(F_i - z_i').$$

The Euler equations are

$$F_{y_j}^* - \frac{d}{dx} F_{y_j'}^* = 0, \quad j = 1, 2, \ldots, n,$$

$$F_{z_i}^* - \frac{d}{dx} F_{z_i'}^* = 0, \quad i = 1, 2, \ldots, m,$$

or

$$F_{y_j} + \sum_{i=1}^{m} \lambda_i F_{i_{y_j}} - \frac{d}{dx} \left(F_{y_j'} + \sum_{i=1}^{m} \lambda_i F_{i_{y_j'}} \right) = 0,$$
$$j = 1, 2, \ldots, n,$$

$$\frac{d}{dx} \lambda_i(x) = 0, \quad i = 1, 2, \ldots, m.$$

It follows from the last m equations, that all λ_i are constant, and the first n equations are the Euler equations for the functional

$$v^{**} = \int\limits_{x_0}^{x_1} \left(F + \sum_{i=1}^{m} \lambda_i F_i \right) dx.$$

We have therefore obtained the following rule: to obtain the fundamental necessary condition for an isoperimetric problem of finding extrema of a functional $v = \int\limits_{x_0}^{x_1} F \, dx$ with constraining relations $\int\limits_{x_0}^{x_1} F_i \, dx = l_i$, $i = 1, 2, \ldots, m$, we have to set up a new functional

$$v^{**} = \int\limits_{x_0}^{x_1} \left(F + \sum_{i=1}^{m} \lambda_i F_i \right) dx,$$

where λ_i are constants, and then write out its Euler equations.

The arbitrary constants $C_1, C_2, \ldots, C_{2n}$ in the general solution of the system of Euler equations and the constants $\lambda_1, \lambda_2, \ldots, \lambda_m$, can be determined from the boundary conditions

$$y_j(x_0) = y_{j0}, \quad y_j(x_1) = y_{j1}, \quad j = 1, 2, \ldots, n$$

and the isoperimetric conditions

$$\int\limits_{x_0}^{x_1} F_i \, dx = l_i, \quad i = 1, 2, \ldots, m.$$

The system of Euler equations for the functional v^{**} remains the same if we multiply v^{**} by a constant multiplier μ_0,

$$\mu_0 v^{**} = \int\limits_{x_0}^{x_1} \sum_{i=0}^{m} \mu_i F_i \, dx,$$

where $F_0 = F$, and $\mu_j = \lambda_j \mu_0$, $j = 1, 2, \ldots, m$. Now all the functions F_i are involved symmetrically, and therefore the extremal curves for the variational problem of finding the extrema of the functional

$$\int_{x_0}^{x_1} F_s\, dx$$

with the isoperimetric conditions

$$\int_{x_0}^{x_1} F_i\, dx = l_i, \quad i = 0, 1, \ldots, s-1, s+1, \ldots, m,$$

and the extremal curves for the original variational problem are the same.

This property is called the *principle of reciprocity*. For instance, the problem of maximum area encircled by a contour curve of given perimeter, and the problem of minimum perimeter of a curve encircling given area are reciprocal and they have their extremals in common.

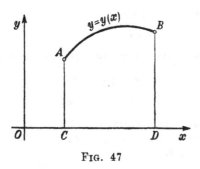

FIG. 47

EXAMPLE 1. *Find a curve of given length l for which the area of a curvilinear trapezoid $CABD$ given in Fig. 47 is maximum.*

Let us investigate the extrema of the functional

$$S = \int_{x_0}^{x_1} y\, dx, \quad y(x_0) = y_0, \quad y(x_1) = y_1,$$

with an isoperimetric condition

$$\int_{x_0}^{x_1} \sqrt{1+y'^2}\,dx = l\,.$$

We set first the auxiliary functional

$$S^{**} = \int_{x_0}^{x_1} (y+\lambda\sqrt{1+y'^2})\,dx\,.$$

Since the integrand does not involve x, it follows that the Euler equation for S^{**} has an integral curve $F - y'F_{y'} = C_1$, or, in this case

$$y + \lambda\sqrt{1+y'^2} - \frac{\lambda y'^2}{\sqrt{1+y'^2}} = C_1\,,$$

and hence

$$y - C_1 = \frac{-\lambda}{\sqrt{1+y'^2}}\,.$$

Introducing a parameter t, setting $y' = \tan t$, we have

$$y - C_1 = -\lambda\cos t, \qquad \frac{dy}{dx} = \tan t,$$

hence

$$dx = \frac{dy}{\tan t} = \frac{\lambda\sin t\,dt}{\tan t} = \lambda\cos t\,dt, \qquad x = \lambda\sin t + C_2\,.$$

Consequently, the parametric equations of extremals are

$$x - C_2 = \lambda\sin t,$$

$$y - C_1 = -\lambda\cos t.$$

Eliminating t, we obtain $(x - C_2)^2 + (y - C_1)^2 = \lambda^2$ which is a family of circles. The constants C_1, C_2, and λ can be determined by the conditions

$$y(x_0) = y_0, \quad y(x_1) = y_1, \quad \text{and} \quad \int_{x_0}^{x_1} \sqrt{1+y'^2}\,dx = l\,.$$

EXAMPLE 2. *Find a curve AB of given length l, which together with a given curve $y = f(x)$ enclose maximum area* (Fig. 48).

We have to determine extrema of the functional

$$S = \int_{x_0}^{x_1} (y - f(x))\,dx, \quad y(x_0) = y_0, \quad y(x_1) = y_1,$$

with the condition that

$$\int_{x_0}^{x_1} \sqrt{1+y'^2}\,dx = l.$$

We set auxiliary functional

$$S^{**} = \int_{x_0}^{x_1} (y - f(x) + \lambda\sqrt{1+y'^2})\,dx.$$

The Euler equation for this functional is the same as in the preceding example, and therefore a maximum may occur only along an arc of a circle.

EXAMPLE 3. *Find the shape of a perfectly flexible unstretchable rope of length l extended between two fixed points A and B* (Fig. 49).

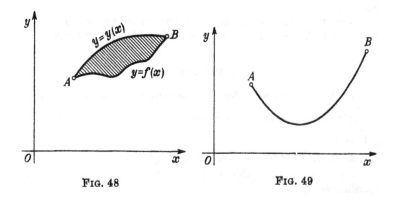

FIG. 48   FIG. 49

Since, in the rest position, the centre of gravity takes the lowest possible position, the problem reduces to that of finding minima of static moment P with respect to x-axis which is directed horizontally. Let us examine the extrema of the functional

$$P = \int_{x_0}^{x_1} y\sqrt{1+y'^2}\,dx$$

with the condition that

$$\int_{x_0}^{x_1} \sqrt{1+y'^2}\,dx = l.$$

We set the auxiliary functional

$$P^{**} = \int_{x_0}^{x_1} (y+\lambda)\sqrt{1+y'^2}\,dx.$$

The Euler equation of this functional has a first integral

$$F - y'F_{y'} = C,$$

or, in this case,

$$(y+\lambda)\sqrt{1+y'^2} - \frac{(y+\lambda)y'^2}{\sqrt{1+y'^2}} = C_1,$$

and so $y+\lambda = C_1\sqrt{1+y'^2}$. Introducing a parameter by setting $y' = \sinh t$, we have $\sqrt{1+y'^2} = \cosh t$, $y+\lambda = C_1\cosh t$, $dy/dx = \sinh t$, $dx = dy/\sinh t = C_1\,dt$, $x = C_1 t + C_2$, or eliminating t, we obtain $y+\lambda = C_1\cosh((x-C_2)/C_1)$, which is a family of catenaries.

The rule for solving isoperimetric problems, given in this section, can be extended to more complicated types of functionals.

Problems

1. Find extremals for an isoperimetric problem $v(y(x)) = \int_0^1 (y'^2 + x^2)\,dx$ with the conditions that $\int_0^1 y^2\,dx = 2$, $y(0) = 0$, $y(1) = 0$.

2. Find geodesics of the circular cylinder $r = R$.

Hint: it is preferable to handle this problem in cylindrical coordinates r, φ, z.

3. Find extremals for an isoperimetric problem

$$v(y(x)) = \int_{x_0}^{x_1} y'^2\,dx$$

with the condition that $\int_{x_0}^{x_1} y\,dx = a$, where a is a constant.

4. Write down the differential equations of extremals for an isoperimetric problem of extrema of the functional

$$v(y(x)) = \int_0^{x_1} (p(x)y'^2 + q(x)y^2)\,dx$$

with the conditions that

$$\int_0^{x_1} r(x)\,y^2\,dx = 1, \quad y(0) = 0, \quad y(x_1) = 0.$$

5. Find the extremal of an isoperimetric problem of extrema of the functional

$$v(y(x),\,z(x)) = \int_0^1 (y'^2 + z'^2 - 4xz' - 4z)\,dx$$

with the conditions:

$$\int_0^1 (y'^2 - xy' - z'^2)\,dx = 2, \quad y(0) = 0, \quad z(0) = 0,$$

$$y(1) = 1, \; z(1) = 1.$$

CHAPTER V

DIRECT METHODS OF SOLVING VARIATIONAL PROBLEMS

1. Direct methods

Differential equations of variational problems can be integrated easily only in exceptional cases. It is therefore essential to find other methods of solution of these problems. The fundamental idea underlying the so called direct methods is to consider a variational problem as a limit problem for some problem of extrema of a function of a finite number of variables. This problem of extrema of a function of a finite number of variables is solved by usual methods, and then by a kind of limiting process the solution of the original variational problem is obtained.

The functional $v\{y(x)\}$ can be considered as a function of an infinite set of variables. This is fairly evident, if we assume that the admissible functions can be represented by a power series

$$y(x) = a_0 + a_1 x + a_2 x^2 + \ldots + a_n x^n + \ldots$$

or by a Fourier series

$$y(x) = \frac{a_2}{2} + \sum_{n=1}^{\infty} (a_n \cos nx + b_n \sin nx)$$

or by any series of the form

$$(1) \qquad y(x) = \sum_{n=1}^{\infty} a_n \varphi_n(x),$$

where $\varphi_n(x)$ are given functions.

To determine a function $y(x)$, which is representable in the form (1), it is sufficient to determine all the coefficients a_n, so that the value of the functional $v(y(x))$ is completely determined by an infinite sequence of numbers $a_1, a_2, \ldots, a_n, \ldots$, i.e. the functional is a function of an infinite set of variables

$$v(y(x)) = \varphi(a_0, a_1, a_2, \ldots, a_n, \ldots).$$

Consequently, the difference between variational problems and problems of extrema of functions of finite number of variables is in the number of variables—in variational problems of extrema this number is infinite. Therefore the fundamental idea of direct methods that consists in considering a variational problem as a limiting case of a problem of extrema of an ordinary function of finite number of variables is of much interest.

In the first period of his investigation of the calculus of variation, L. Euler applied a method which is now called the direct method of finite differences. This method has not been applied since then, and it was only in the last three decades when it was revived with much success by Soviet mathematicians (L. A. Lusternik, I. G. Petrovski and others).

Another direct method is a method of Ritz which has been greatly developed by Soviet mathematicians (N. M. Krylov, N. N. Bogolubov and others). Nowadays it is widely used to solve various variational problems, in particular those encountered in the theory of elasticity.

A third method of L. V. Kantorovič applies to functionals which depend on functions of several independent variables, and it is used more and more widely in those domains where the Ritz method is used.

In the sequel, we shall consider only these three fundamental methods, and many assertions will be given without proof. The reader who wishes to know more details about contemporary direct methods now in use, is referred to a book by L. V. Kantorovič, V. I. Krylov, *Approximate methods of higher analysis* (in Russian),

Moscow 1952, and a book of S. G. Mikhlin, *Direct methods of mathematical physics* (in Russian), Moscow 1950.

2. Euler method of finite differences

The main idea of the *method of finite differences* is that the values of a functional $v(y(x))$, say

$$\int_{x_0}^{x_1} F(x, y, y')\, dx, \quad y(x_0) = a, \quad y(x_1) = b,$$

are considered not along arbitrary admitted for a given variational problem curves, but only along polygonal curves which consist of a prescribed number of line segments with fixed abscissae of vertices: $x_0 + \Delta x, x_0 + 2\Delta x, \ldots,$ $x_0 + (n-1)\Delta x$, where $\Delta x = (x_1 - x_0)/n$ (Fig. 50). Along such polygonal curves the functional $v(y(x))$ turns into a function $\varphi(y_1, y_2, \ldots, y_{n-1})$ of ordinates $y_1, y_2, \ldots, y_{n-1}$ of vertices, for a polygonal curve is completely determined by these ordinates.

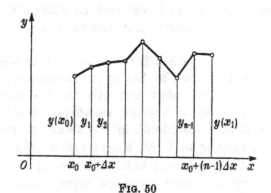

FIG. 50

We will choose the ordinates $y_1, y_2, \ldots, y_{n-1}$ so that the function $\varphi(y_1, y_2, \ldots, y_{n-1})$ has an extremum, i.e. we determine $y_1, y_2, \ldots, y_{n-1}$ by the equations

$$\frac{\partial \varphi}{\partial y_1} = 0, \quad \frac{\partial \varphi}{\partial y_2} = 0, \quad \ldots, \quad \frac{\partial \varphi}{\partial y_{n-1}} = 0,$$

and then pass to the limit with $n \to \infty$. With some restrictions imposed on the function F, we shall obtain in this way the solution of the original variational problem.

It is, however, more convenient to calculate the value of the functional $v(y(x))$ along those polygonal curves just approximately, for instance in the case of the simplest problem, it is convenient to replace the integral

$$\int_{x_0}^{x_1} F(x, y, y') dx = \sum_{k=0}^{n-1} \int_{x_0+k\Delta x}^{x_0+(k+1)\Delta x} F\left(x, y, \frac{y_{k+1}-y_k}{\Delta x}\right) dx$$

by a finite sum

$$\sum_{i=1}^{n} F\left(x_i, y_i, \frac{\Delta y_i}{\Delta x}\right) \Delta x.$$

To illustrate this method, we shall derive the Euler equations for the functional

$$v(y(x)) = \int_{x_0}^{x_1} F(x, y, y') dx.$$

In this case we have

$$v(y(x)) \approx \varphi(y_1, y_2, \ldots, y_{n-1}) = \sum_{i=0}^{n-1} F\left(x_i, y_i, \frac{y_{i+1}-y_i}{\Delta x}\right) \Delta x$$

along the polygonal curves. Since there are only two terms in this sum that depend on y_i, namely, the i-th term and $(i-1)$-th term,

$$F\left(x_i, y_i, \frac{y_{i+1}-y_i}{\Delta x}\right) \Delta x \quad \text{and} \quad F\left(x_{i-1}, y_{i-1}, \frac{y_i-y_{i-1}}{\Delta x}\right) \Delta x,$$

it follows that the equations $\partial\varphi/\partial y_i = 0$, $i = 1, 2, \ldots, n-1$, take the form

$$F_y\left(x_i, y_i, \frac{y_{i+1} - y_i}{\Delta x}\right)\Delta x + F_{y'}\left(x_i, y_i, \frac{y_{i+1} - y_i}{\Delta x}\right)\left(-\frac{1}{\Delta x}\right)\Delta x +$$

$$+ F_{y'}\left(x_{i-1}, y_{i-1}, \frac{y_i - y_{i-1}}{\Delta x}\right)\frac{1}{\Delta x}\Delta x = 0,$$

$$i = 1, 2, \ldots, n-1,$$

or

$$F_y\left(x_i, y_i, \frac{\Delta y_i}{\Delta x}\right) -$$

$$- \frac{F_{y'}(x_i, y_i, \Delta y_i/\Delta x) - F_{y'}(x_{i-1}, y_{i-1}, \Delta y_{i-1}/\Delta x)}{\Delta x} = 0,$$

or

$$F_y\left(x_i, y_i, \frac{\Delta y_i}{\Delta x}\right) - \frac{\Delta F_{y'}}{\Delta x} = 0,$$

hence letting $n \to \infty$ we obtain Euler equations

$$F_y - \frac{d}{dx} F_{y'} = 0,$$

which must be satisfied by the function $y(x)$ giving an extremum. The fundamental necessary conditions of extremum for other variational problems can be obtained in a similar way.

If we do not wish to complete the limiting process in full, we can determine from the equations $\partial\varphi/\partial y_i$, $i = 1, 2, \ldots, n-1$, the ordinates $y_1, y_2, \ldots, y_{n-1}$ and by so doing we shall have a polygonal curve which is an approximate solution to the variational problem in question.

3. Ritz's method

The idea of *Ritz's method* is to consider a functional $v(y(x))$ not along arbitrary admissible curves, but only along all possible linear combinations

$$y_n = \sum_{i=1}^{n} a_i W_i(x),$$

with constant coefficients, of the first n functions of a certain sequence of functions

$$W_1(x),\ W_2(x),\ \ldots,\ W_n(x),\ \ldots$$

The functions

$$y_n = \sum_{i=1}^{n} a_i W_i(x)$$

should be admissible functions for a given problem, which is an additional restriction on the choice of the sequence of functions $W_i(x)$. Along such linear combinations the functional $v(y(x))$ becomes a function $\varphi(a_1, a_2, \ldots, a_n)$ of the coefficients $a_1, a_2, \ldots, a_n$. These coefficients $a_1, a_2, \ldots, a_n$ are then so chosen that the function $\varphi(a_1, \ldots, a_n)$ has an extremum. Consequently, $a_1, a_2, \ldots, a_n$ shall be determined by the system of equations

$$\frac{\partial \varphi}{\partial a_i} = 0, \quad i = 1, 2, \ldots, n.$$

Completing the limiting process by letting $n \to \infty$, we shall obtain a limit function

$$y = \sum_{i=1}^{\infty} a_i W_i(x),$$

provided this series converges. Under some assumptions about the functional $v(y(x))$ and the sequence $W_1(x)$,

$W_2(x), \ldots, W_n(x), \ldots$, this function will be an exact solution of the variational problem in question. If we do not wish to carry out the limiting process in full, and would rather be satisfied with the first n terms

$$y_n = \sum_{i=1}^{n} a_i W_i(x)$$

we will have an approximate solution of the variational problem.

If we determine a minimum of a functional in this way, then the exact value of this minimum is approximated from above, for the minimum of the functional extended over wider class of all admissible curves is always less or equal to the minimum of this same functional extended only over a part of these curves — those of the form $y_n = \sum_{i=1}^{n} a_i W_i(x)$. By the same token, this method used to determine a maximum value of a functional will yield approximations from below.

In order that

$$y_n = \sum_{i=1}^{n} a_i W_i(x)$$

be admissible functions, they must satisfy the boundary conditions, and of course, some other conditions like continuity, smoothness etc. If the boundary conditions are linear and homogeneous, for instance, in the case of a simplest problem $y(x_0) = y(x_1) = 0$, or

$$\beta_{1j} y(x_j) + \beta_{2j} y'(x_j) = 0, \quad j = 0, 1,$$

where β_{ij} are constants, then it is most convenient to choose the coordinate functions $W_n(x)$ in such a way that they all satisfy these boundary conditions. This being done, it is obvious that also $y_n = \sum_{i=1}^{n} a_i W_i(x)$ for arbitrary a_i will satisfy these same boundary conditions.

For instance, suppose that the boundary conditions are $y(x_0) = y(x_1) = 0$. Then we can choose

$$W_i(x) = (x - x_0)(x - x_1)\varphi_i(x)$$

as coordinate functions, where $\varphi_i(x)$ are arbitrary continuous functions, or

$$W_k(x) = \sin\frac{k\pi(x - x_0)}{x_1 - x_0}, \quad k = 1, 2, \ldots$$

or any other functions satisfying the conditions

$$W_i(x_0) = W_i(x_1) = 0.$$

If the boundary conditions are not homogeneous, for instance $y(x_0) = y_0$, $y(x_1) = y_1$, where at least one of the numbers y_0 or y_1 is different from zero, then the most convenient way is to try to find a solution of the variational problem in the form

$$y_n = \sum_{i=1}^{n} a_i W_i(x) + W_0(x),$$

where $W_0(x)$ satisfies the given boundary conditions $W_0(x_0) = y_0$, $W_0(x_1) = y_1$, and all the remaining functions $W_i(x)$ satisfy the corresponding homogeneous boundary conditions, i. e. in this case $W_i(x_0) = W_i(x_1) = 0$. The sequence $W_i(x)$ being so chosen, $y_n(x)$ satisfy the given non-homogeneous boundary conditions for arbitrary a_i. As the function $W_0(x)$, we can take, for instance, a linear function

$$W_0(x) = \frac{y_1 - y_0}{x_1 - x_0}(x - x_0) + y_0.$$

To solve the system of equations $\partial\varphi/\partial a_i = 0$, $i = 1, 2, \ldots, n$, is in general a hard task. This problem is much simplified, when the functional v investigated is a quadratic functional with respect to unknown functions and its derivatives, for in this case the equations $\partial\varphi/\partial a_i = 0$, $i = 1, 2, \ldots, n$, are linear with respect to a_i.

The initial choice of the functions W_1, W_2, ..., W_n, ..., which are called *coordinate functions*, has much influence on the degree of complexity of the subsequent computation, and therefore a successful application of Ritz's method to a great extent depends on an adequate choice of the coordinate functions.

All this applies in full to the functionals $v(z(x_1, x_2, \dots \dots, x_n))$, but of course then the functions W_i must be functions of the variables $x_1, x_2, \dots, x_n$. The same holds of functionals depending on several functions.

Ritz's method is often used to find an exact or approximate solution of problems of mathematical physics. For instance, if we wish to find the solution of Poisson's equation

$$\frac{\partial^2 z}{\partial x^2} + \frac{\partial^2 z}{\partial y^2} = f(x, y)$$

in some domain D, where the values of z on the boundary of D are given, we can replace this problem by a variational problem of extrema of a functional for which the given equation is an Ostrogradski equation (cf. p. 50). In the case of Poisson's equation this functional is

$$\iint_D \left(\left(\frac{\partial z}{\partial x}\right)^2 + \left(\frac{\partial z}{\partial y}\right)^2 + 2zf(x, y) \right) dx dy .$$

The functions z that make this functional have an extremum can be investigated by any of the direct methods.

The problems of mathematical physics can generally be reduced to problems of extrema of functionals that are second-order with respect to unknown functions and their derivatives and therefore the application of Ritz's method in such cases is a simpler task.

The question of convergence of Ritz approximations to the desired solution of a variational problem, as well as the problem of estimation of the degree of approximation is very complicated. Therefore we shall make only a few remarks, and advise the reader who wishes to know more on the subject to refer to the books of Mikhlin, Kantorovič and Krylov cited on p. 149-150.

To make the discussion more specific we will direct our attention to the functional

$$v(y(x)) = \int_{x_0}^{x_1} F(x, y(x), y'(x)) dx$$

and assume that the extremum in question is a minimum. The sequence of the coordinate functions $W_1(x)$, $W_2(x)$, ..., $W_n(x)$, ... is assumed to be complete in the sense that each admissible function can be approximated arbitrarily first-order close by linear combinations

$$\sum_{k=1}^{n} a_k W_k(x)$$

of coordinate functions, where n is sufficiently large. It is evident that we can construct by Ritz's method a sequence of functions $y_1, y_2, \ldots, y_n, \ldots$, where

$$y_n = \sum_{k=1}^{n} a_k W_k(x),$$

which is a so called *minimizing sequence*, i.e. one for which the values of the functional

$$v(y_1), v(y_2), \ldots, v(y_n)$$

tend to the minimum value, or to the greatest lower bound of the functional $v(y(x))$. However, it does not follow from the relation $\lim_{n\to\infty} v(y_n(x)) = \min v(y(x))$ that $\lim_{n\to\infty} y_n(x)$ makes v a minimum. A minimizing sequence may not tend to a function giving an extremum in the class of admissible functions.

In fact, the difference between the values

$$v\big(y_n(x)\big) = \int_{x_0}^{x_1} F\big(x, y_n(x), y_n'(x)\big)\,dx$$

and

$$v\big(y(x)\big) = \int_{x_0}^{x_1} F\big(x, y(x), y'(x)\big)\,dx$$

of the functional may be very small not only in the case when $y_n(x)$ is first-order close to $y(x)$ on the whole interval of integration (x_0, x_1), but also in a case when in sufficiently small parts of the interval (x_0, x_1) the functions $y_n(x), y(x)$ and their derivatives differ considerably, while on the other parts of this interval they are close to each other (Fig. 51). Consequently a minimizing se-

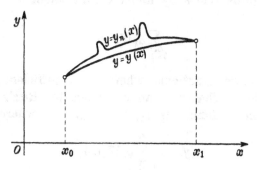

Fig. 51

quence $y_1, y_2, \ldots, y_n, \ldots$ may not have a limit in the class of admissible functions, even though the functions $y_1, y_2, \ldots, y_n, \ldots$ themselves are admissible functions.

The conditions for convergence of a sequence y_n, constructed by the method of Ritz to a solution of a variational problem and an estimate of the rate of convergence for particular functionals of the kinds most frequently occurring were developed in detail in papers by N. M. Krylov and N. N. Bogolubov, for instance, for a functional of the form

$$v = \int\limits_0^1 \left(p(x)y'^2 + q(x)y^2 + f(x)y\right) dx, \quad y(0) = y(1) = 0,$$

where $p(x) > 0$, $q(x) \geqslant 0$, which is often encountered in applied problems. The coordinate functions being

$$W_k(x) = \sqrt{2}\sin k\pi x, \quad k = 1, 2, \ldots,$$

it was not only proved that the Ritz's approximations to the function $y(x)$ giving a minimum of the functional converge but also very close estimates of the error $|y(x) - y_n(x)|$ were derived.

We shall give one of them. Although it is not the best one, it is a very convenient estimate of the maximum of $|y(x) - y_n(x)|$ on the interval $(0, 1)$:

$$\max |y - y_n|$$

$$\leqslant \frac{1}{n+1}\left(\max p(x) + \frac{\max q(x)}{(n+1)^2\pi^2)}\right)^{1/2} \frac{\sqrt{\int\limits_0^1 f^2(x)\,dx}}{\pi^2\sqrt{2}\left(\min p(x)\right)^{5/2}} \times$$

$$\times \left(\max |p'(x)| + \frac{1}{\pi}\max q(x) + \pi \min p(x)\right)(^1).$$

Even in this comparatively simple case the estimate of error is very complicated. Therefore, to determine the exactness of the results obtained by Ritz's method or other direct methods, we usually use the following practical method, which does not however have an adequate theoretical foundation. Having computed $y_n(x)$ and $y_{n+1}(x)$, we compare them at some points of the interval (x_0, x_1). If with the degree of exactness required for a given purpose their values coincide, then we consider the solution of the variational problem equal to $y_n(x)$. When, for some values of x, $y_n(x)$ and $y_{n+1}(x)$ differ too much for the given purpose, then we compute $y_{n+2}(x)$ and compare $y_{n+1}(x)$ and $y_{n+2}(x)$ at some points of the interval (x_0, x_1). This

---

(¹) Cf. Kantorovič and Krylov *loc. cit.*

process is continued until $y_{n+k}(x)$ and $y_{n+k+1}(x)$ can be considered for the same the given purpose.

EXAMPLE 1. *Investigating the vibrations of a fixed wedge of constant thickness* (Fig. 52), *we have to examine the extrema of the functional*

$$v = \int_0^1 (ax^3 y''^2 - bxy^2)\, dx, \quad y(1) = y'(1) = 0,$$

where a and b are positive constants.

As coordinate functions satisfying these boundary conditions we can take

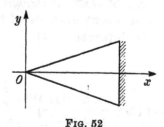

FIG. 52

$$(x-1)^2, \quad (x-1)^2 x, \quad (x-1)^2 x^2, \quad \dots, \quad (x-1)^2 x^{k-1}, \quad \dots,$$

and consequently

$$y_n = \sum_{k=1}^n a_k (x-1)^2 x^{k-1}.$$

Taking only the first two terms, we have

$$y_2 = (x-1)^2 (a_1 + a_2 x)$$

so that

$$v_2 = v(y_2) = \int_0^1 (ax^3 (6a_2 x + 2a_1 - 4a_2)^2 - bx(x-1)^4 (x-1)(a_1 + a_2 x)^2)\, dx$$

$$= a\left((a_1 - 2a_2)^2 + \frac{24}{5} a_2 (a_1 - 2a_2) + 6a_2^2\right) -$$

$$- b\left(\frac{a_1^2}{30} + \frac{2a_1 a_2}{105} + \frac{a_2^2}{280}\right).$$

The necessary condition of an extremum $\partial v_2 / \partial a_1 = 0$, $\partial v_2 / \partial a_2 = 0$, takes the form

$$\left(a - \frac{b}{30}\right) a_1 + \left(\frac{2}{5} a - \frac{b}{105}\right) a_2 = 0$$

or

$$\left(\frac{2}{5} a - \frac{b}{105}\right) a_1 + \left(\frac{2}{5} a - \frac{b}{280}\right) a_2 = 0.$$

In order to obtain solutions that are different from the zero solution $a_1 = a_2 = 0$, which corresponds to the equilibrium position with no vibrations at all, it is necessary that the determinant of this linear system of equations be zero,

$$\begin{vmatrix} a - \dfrac{b}{30} & \dfrac{2}{5} a - \dfrac{b}{105} \\[2ex] \dfrac{2}{5} a - \dfrac{b}{105} & \dfrac{2}{5} a - \dfrac{b}{280} \end{vmatrix} = 0$$

or

$$\left(a - \frac{b}{30}\right)\left(\frac{2}{5} a - \frac{b}{280}\right) - \left(\frac{2}{5} a - \frac{b}{105}\right)^2 = 0.$$

This equation is called the *frequency equation*. It determines the frequency b of free vibrations of the wedge described by the function

$$u(x, t) = y(x) \cos bt.$$

The frequency equation has in general two roots b_1 and b_2. The smaller of them gives an approximate value of the fundamental tone of vibration of the wedge.

EXAMPLE 2. *In problems connected with the torsion of a prism or cylinder, we have to examine the extrema of the functional*

$$v(z(x, y)) = \iint\limits_{D} \left(\left(\frac{\partial z}{\partial x} - y\right)^2 + \left(\frac{\partial z}{\partial y} + x\right)^2\right) dx\, dy.$$

In the case of a cylinder with an elliptic base the domain of integration D is the interior of the ellipse

$$\frac{x^2}{a^2} + \frac{y^2}{b^2} = 1.$$

In this case, taking only one coordinate function, xy, we have

$$z_1 = axy, \qquad v(z_1) = v_1 = \frac{\pi ab}{4}\left((a+1)^2 a^2 + (a-1)^2 b^2\right).$$

The necessary condition of an extremum $\partial v_1/\partial a = 0$ turns into

$$(a+1)a^2 + (a-1)b^2 = 0,$$

and consequently

$$a = \frac{b^2 - a^2}{b^2 + a^2}, \qquad z_1 = \frac{b^2 - a^2}{b^2 + a^2}\,xy.$$

EXAMPLE 3. *With the assumptions of the preceding example, find z assuming in addition that the domain D is a rectangle with sides $2a$ and $2b$, $-a \leqslant x \leqslant a$, $-b \leqslant y \leqslant b$.*

Taking as the coordinate functions xy, xy^3, x^3y, i.e. setting

$$z_3 = a_1 xy + a_2 xy^3 + a_3 x^3 y,$$

we have

$$v_3 = v(z_3) = \int\limits_{-a}^{a}\int\limits_{-b}^{b}\left(\left(\frac{\partial z_3}{\partial x} - y\right)^2 + \left(\frac{\partial z_3}{\partial y} + x\right)^2\right) dx\,dy$$

$$= \frac{4}{3}ab^3(a_1 - 1)^2 + 4ab^5\left(\frac{b^2}{7} + \frac{3a^2}{5}\right)a_2^2 + 4a^5 b\left(\frac{a^2}{7} + \frac{3b^2}{5}\right)a_3^2 +$$

$$+ \frac{4}{3}a^3 b(a_1 + 1)^2 + \frac{8}{5}ab^5(a_1 - 1)a_2 + \frac{8}{5}a^3 b(a_1 + 1)a_2 -$$

$$- \frac{8}{5}a^5 b(a_1 + 1)a_3 - \frac{8}{5}a^3 b^3(a^2 + b^2)a_2 a_3 - \frac{8}{3}a^3 b^3(a_1 - 1)a .$$

The fundamental necessary conditions of an extremum

$$\frac{\partial v_3}{\partial a_1} = 0, \qquad \frac{\partial v_3}{\partial a_2} = 0, \qquad \frac{\partial v_3}{\partial a_3} = 0$$

enable us to calculate a_1, a_2, a_3,

$$a_1 = -\frac{7(a^6 - b^6) + 135a^2 b^2 (a^2 - b^2)}{7(a^6 + b^6) + 107a^2 b^2 (a^2 + b^2)},$$

$$a_2 = -\frac{7a^2 (3a^2 + 35b^2)}{21(a^6 + b^6) + 321a^2 b^2 (a^2 + b^2)},$$

$$a_3 = -\frac{7b^2 (35a^2 + 3b^2)}{21(a^6 + b^6) + 321a^2 b^2 (a^2 + b^2)}.$$

EXAMPLE 4. *Find the solution of the equation*

$$\frac{\partial^2 z}{\partial x^2} + \frac{\partial^2 z}{\partial y^2} = f(x, y)$$

in the interior of a rectangle D, $0 \leqslant x \leqslant a$, $0 \leqslant y \leqslant b$, with the assumption that this solution shall vanish on the boundary of D.

We assume that the function $f(x, y)$ can be decomposed into a double Fourier series

$$f(x, y) = \sum_{p=1}^{\infty} \sum_{q=1}^{\infty} \beta_{pq} \sin p \frac{\pi x}{a} \sin q \frac{\pi y}{b},$$

which converges to $f(x, y)$ uniformly in the interior of D. This boundary problem can be reduced to a variational problem, i.e. we can choose a functional, for which the given equation is the Ostrogradski equation, and then by means of one of the direct methods we can find the solution of the original boundary problem. It is easy to see that

$$\frac{\partial^2 z}{\partial x^2} + \frac{\partial^2 z}{\partial y^2} = f(x, y)$$

is the Ostrogradski equation for the functional

$$v(z(x, y)) = \iint\limits_{D} \left(\left(\frac{\partial z}{\partial x}\right)^2 + \left(\frac{\partial z}{\partial y}\right)^2 + 2zf(x, y) \right) dx dy$$

(cf. p. 51). The boundary condition for the functional is the same: $z = 0$ on the boundary of the domain D. Let us examine the extrema of this functional by Ritz's method. We take

$$\sin m \frac{\pi x}{a} \sin n \frac{\pi y}{b}, \quad m, n = 1, 2, \ldots,$$

as coordinate functions. Each of these functions satisfies the boundary condition $z = 0$ on the boundary of D, as well as all the linear combinations of them. Moreover, this system of functions is complete. Setting

$$z_{nm} = \sum_{p=1}^{n} \sum_{q=1}^{m} a_{pq} \sin p \frac{\pi x}{a} \sin q \frac{\pi y}{b}$$

we have

$$v(z_{nm}) = \int_0^a \int_0^b \left(\left(\frac{\partial z_{nm}}{\partial x} \right)^2 + \left(\frac{\partial z_{nm}}{\partial y} \right)^2 + \right.$$

$$\left. + 2 z_{nm} \sum_{p=1}^{\infty} \sum_{q=1}^{\infty} \beta_{pq} \sin p \frac{\pi x}{a} \sin q \frac{\pi y}{b} \right) dx dy$$

$$= \frac{\pi^2 ab}{4} \sum_{p=1}^{n} \sum_{q=1}^{m} \left(\frac{p^2}{a^2} + \frac{q^2}{b^2} \right) a_{pq}^2 + \frac{ab}{2} \sum_{p=1}^{n} \sum_{q=1}^{m} a_{pq} \beta_{pq}.$$

This is easy to obtain, if we observe that the coordinate functions $\sin p \dfrac{\pi x}{a} \sin q \dfrac{\pi y}{b}$ $(p, q = 1, 2, \ldots)$ form an orthogonal system in the domain D, i. e.

$$\iint_D \sin p \frac{\pi x}{a} \sin q \frac{\pi y}{b} \sin p_1 \frac{\pi x}{a} \sin q_1 \frac{\pi y}{b} dx dy = 0$$

for arbitrary positive integers p, q, p_1, q_1, with the exception of such cases as $p = p_1, q = q_1$. If $p = p_1$ and $q = q_1$, we have

$$\iint_D \sin^2 p \frac{\pi x}{a} \sin^2 q \frac{\pi y}{b} dx dy = \frac{ab}{4}.$$

Therefore, in the integral expression of $v(z_{nm})$, only such terms of the integrand shall be taken which involve the squares of the functions

$$\sin p \frac{\pi x}{a} \sin q \frac{\pi y}{b}, \qquad \sin p \frac{\pi x}{a} \cos q \frac{\pi y}{b}, \qquad \cos p \frac{\pi x}{a} \sin q \frac{\pi y}{b}.$$

It is obvious that $v(z_{nm})$ is a function $\varphi(a_{11}, a_{12}, \ldots, a_{nm})$ of the coefficients $a_{11}, a_{12}, \ldots, a_{nm}$, which can be determined by the fundamental necessary condition of an extremum

$$\frac{\partial \varphi}{\partial a_{pq}} = 0, \quad p = 1, 2, \ldots, n; \quad q = 1, 2, \ldots, m.$$

This system of equations takes the form

$$a_{pq} \left(\frac{p^2}{a^2} + \frac{q_2}{b^2} \right) \pi^2 + \beta_{pq} = 0, \quad p = 1, 2, \ldots, n, \quad q = 1, 2, \ldots, m,$$

and hence

$$a_{pq} = -\frac{\beta_{pq}}{\pi^2\left(\dfrac{p^2}{a^2} + \dfrac{q^2}{b^2}\right)}.$$

Consequently

$$z_{nm} = -\frac{1}{\pi^2}\sum_{p=1}^{n}\sum_{q=1}^{m}\frac{\beta_{pq}}{\dfrac{p^2}{a^2} + \dfrac{q^2}{b^2}}\sin p\,\frac{\pi x}{a}\sin q\,\frac{\pi y}{b}.$$

By a limiting process, as n and m both increase to infinity, we obtain in this particular case the exact solution

$$z = -\frac{1}{\pi^2}\sum_{p=1}^{\infty}\sum_{q=1}^{\infty}\frac{\beta_{pq}}{\dfrac{p^2}{a^2} + \dfrac{q^2}{b^2}}\sin p\,\frac{\pi x}{a}\sin q\,\frac{\pi y}{b}.$$

4. Kantorovič's method

When applying Ritz's method to functionals of the form $v\big(z(x_1, x_2, \ldots, x_n)\big)$, depending on functions of several independent variables, we choose a system of coordinate functions

$$W_1(x_1, x_2, \ldots, x_n),\ W_2(x_1, x_2, \ldots, x_n),\ \ldots,$$
$$W_m(x_1, x_2, \ldots, x_n),\ldots$$

and an approximate solution of the given variational problem is to be determined in the form

$$z_m = \sum_{k=1}^{m} a_k W_k(x_1, \ldots, x_n),$$

where the coefficients a_k are constant.

The *Kantorovič method* consists also in choosing a system of coordinate functions

$$W_1(x_1, x_2, \ldots, x_n),\ W_2(x_1, x_2, \ldots, x_n),\ \ldots,$$
$$W_m(x_1, x_2, \ldots, x_n),\ldots$$

but then we try to determine an approximate solution of the form

$$z_m = \sum_{k=1}^{m} a_k(x_i) W_k(x_1, x_2, \ldots, x_n),$$

where the coefficients $a_k(x_i)$ are no longer constants, but are unknown functions of one of the independent variables. The functional $v(z)$ considered only for functions of the form

$$z_m = \sum_{k=1}^{m} a_k(x) W_k(x_1, x_2, \ldots, x_n)$$

turns into a functional $\tilde{v}\big(a_1(x_i), a_2(x_i), \ldots, a_m(x_i)\big)$, depending on m functions of one independent variable

$$a_1(x_i), a_2(x_i), \ldots, a_m(x_i).$$

These functions must be so chosen that the functional $\tilde{v}$ has an extremum.

This being done, by a limiting process with $m \to \infty$, we may, in certain conditions, obtain the exact solution. If the limiting process is not completed, we shall obtain in this way an approximate solution, which, in general, is far better than that obtained by Ritz's method with the same coordinate functions and the same number of terms m.

The reason why approximations obtained this way are better is that the class of functions

$$z_m = \sum_{k=1}^{m} a_k(x_i) W_k(x_1, x_2, \ldots, x_n),$$

where $a_k(x_i)$ are not constants but functions of one variable x_i, is much wider than the class of functions

$$z_m = \sum_{k=1}^{m} a_k W_k(x_1, x_2, \ldots, x_n),$$

where a_k are constants.

For instance, let us suppose that we have to find the extrema of the functional

$$v = \int\limits_{x_0}^{x_1} \int\limits_{\varphi_1(x)}^{\varphi_2(x)} F\left(x, y, z, \frac{\partial z}{\partial x}, \frac{\partial z}{\partial y}\right) dx dy,$$

where the integration is over a domain D determined by the curves $y = \varphi_1(x)$, $y = \varphi_2(x)$ and two lines $x = x_0$, $x = x_1$ (Fig. 53). The values of the function $z(x, y)$ on the boundary of D are given. We choose a sequence of coordinate functions

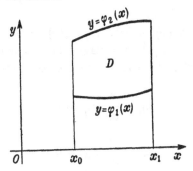

FIG. 53

$$W_1(x, y), W_2(x, y), \ldots, W_n(x, y), \ldots$$

If we take only the first m functions of this sequence, we will try to find a solution of the variational problem in the form

$$z_m = \sum_{k=1}^{m} a_k(x) W_k(x, y),$$

or replacing $a_k(x)$ by $u_k(x)$, we have

$$z_m(x, y) = u_1(x) W_1(x, y) + u_2(x) W_2(x, y) + \ldots + \\ + u_m(x) W_m(x, y),$$

where W_k are the just chosen coordinate functions, and u_k are unknown functions, which shall be so determined that they make v have an extremum.

We have

$$v\big(z_m(x, y)\big) = \int\limits_{x_0}^{x_1} dx \int\limits_{\varphi_1(x)}^{\varphi_2(x)} F\left(x, y, z_m(x, y), \frac{\partial z_m}{\partial x}, \frac{\partial z_m}{\partial y}\right) dy.$$

Since the integrand function is known and depends only on y, the integration with respect to y can be performed, so that the functional $v\big(z_m(x, y)\big)$ would be of the form

$$v\big(z_m(x, y)\big) = \int\limits_{x_0}^{x_1} \varphi\big(x, u_1(x), \ldots, u_m(x), u_1', \ldots, u_m'\big) dx.$$

The functions $u_1(x), u_2(x), \ldots, u_m(x)$ shall be chosen so as to make $v\big(z_m(x, y)\big)$ an extremum. Consequently $u_i(x)$ shall satisfy the following system of Euler equations

$$\varphi_{u_1} - \frac{d}{dx}\varphi_{u_1'} = 0,$$

$$\varphi_{u_2} - \frac{d}{dx}\varphi_{u_2'} = 0,$$

$$\cdots \cdots \cdots \cdots$$

$$\varphi_{u_m} - \frac{d}{dx}\varphi_{u_m'} = 0.$$

The arbitrary constants shall be so chosen that $z_m(x, y)$ satisfy the given boundary conditions on the straight lines $x = x_0$ and $x = x_1$.

EXAMPLE 1. *Examine the extrema of the functional*

$$v\,(z\,(x, y)) = \int\limits_{-a}^{a} \int\limits_{-b}^{b} \left(\left(\frac{\partial z}{\partial x}\right)^2 + \left(\frac{\partial z}{\partial y}\right)^2 - 2z\right) dx\,dy,$$

where $z = 0$ on the boundary of the domain of integration.

This domain is a rectangle $-a \leqslant x \leqslant a$, $-b \leqslant y \leqslant b$. We will try to find a solution of the form $z_1 = (b^2 - y^2)u(x)$. Of course, functions of this kind satisfy the boundary conditions on the lines

$y = \pm b$. We have

$$v(z_1) = \int\limits_{-a}^{a} \left(\frac{16}{15} b^5 u'^2 + \frac{8}{3} b^3 u^2 - \frac{8}{3} b^3 u \right) dx.$$

The Euler equation for this functional,

$$u'' - \frac{5}{2b^2} u = -\frac{5}{4b^2}$$

is a linear differential equation with constant coefficients. The general solution is

$$u = C_1 \cosh \sqrt{\frac{5}{2}} \cdot \frac{x}{b} + C_2 \sinh \sqrt{\frac{5}{2}} \cdot \frac{x}{b} + \frac{1}{2}.$$

The constants C_1 and C_2 are determined by the boundary conditions $z(-a) = z(a) = 0$. We have

$$C_2 = 0, \quad C_1 = -\frac{1}{2\cosh \sqrt{\dfrac{5}{2}} \cdot \dfrac{a}{b}},$$

hence

$$u = \frac{1}{2} \left(1 - \frac{\cosh \sqrt{\dfrac{5}{2}} \cdot \dfrac{x}{b}}{\cosh \sqrt{\dfrac{5}{2}} \cdot \dfrac{a}{b}} \right)$$

and consequently

$$z_1 = \frac{1}{2} (b^2 - y^2) \left(1 - \frac{\cosh \sqrt{\dfrac{5}{2}} \cdot \dfrac{x}{b}}{\cosh \sqrt{\dfrac{5}{2}} \cdot \dfrac{a}{b}} \right).$$

If we wish to obtain a more exact solution, we can try to find a solution of the form

$$z_2 = (b^2 - y^2) u_1(x) + (b^2 - y^2)^2 u_2(x).$$

EXAMPLE 2. *Find a continuous solution of the equation* $\Delta z = -1$ *in a domain* D, *which is an isosceles triangle determined by the lines* $y = \pm \dfrac{\sqrt{3}}{3} x$, *and* $x = b$ (Fig. 54). *It is required that this solution be zero on the boundary of the domain* D.

The equation $\varDelta z = -1$ is the Ostrogradski equation for the functional

$$v(z) = \int_0^b \int_{-\frac{\sqrt{3}}{3}x}^{\frac{\sqrt{3}}{3}x} \left(\left(\frac{\partial z}{\partial x} \right)^2 + \left(\frac{\partial z}{\partial y} \right)^2 - 2z \right) dx\,dy,$$

with the assumption that $z = 0$ on the boundary of the domain of integration. Following the Kantorovič's method, we will try to find the first approximation in the form

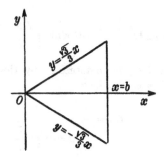

FIG. 54

$$z_1 = \left(y^2 - \left(\frac{\sqrt{3}}{3} x \right)^2 \right) u(x).$$

With z_1 so chosen, the boundary conditions on the lines $y = \pm \dfrac{\sqrt{3}}{3} x$ are satisfied.

The integration with respect to y being performed, the functional $v(z_1)$ takes the form

$$v(z_1) = \frac{8\sqrt{3}}{405} \int_0^b (2x^5 u'^2 + 10x^4 uu' + 30x^3 u^2 + 15x^3 u)\, dx.$$

Its Euler equation is $x^2 u'' + 5xu' - 5u = \dfrac{15}{4}$. In the theory of differential equations, linear equations of this kind are called *Euler equations*.

One particular solution of this non-homogeneous equation is obvious: $u = -\dfrac{3}{4}$. The solution of the corresponding homogeneous equation can be found in the form $u = x^k$, and finally we have $u = C_1 x + C_2 x^{-5} - \frac{3}{4}$. Since in some neighbourhood of the point $x = 0$ the solution u must be bounded, C_2 must vanish. On the other hand, by $u(b) = 0$, we have $C_1 = -3/4b$, and consequently

$$z_1 = -\frac{3}{4}\left(1 - \frac{x}{b}\right)\left(y^2 - \frac{1}{3}x^2\right).$$

Problems

1. Find an approximate solution of the equation $\Delta z = -1$ in the interior of a square $-a \leqslant x \leqslant a$, $-a \leqslant y \leqslant a$, with the boundary condition that this solution vanishes on the boundary of the square.

Hint: This problem reduces to one of examining extrema of the functional

$$\iint_D \left(\left(\frac{\partial z}{\partial x}\right)^2 + \left(\frac{\partial z}{\partial y}\right)^2 - 2z\right) dx\,dy.$$

An approximate solution may be determined in the form

$$z_0 = a(x^2 - a^2)(y^2 - a^2).$$

2. Find an approximate solution of a problem of extrema of the functional

$$v(y(x)) = \int_0^1 (x^3 y''^2 + 100xy^2 - 20xy)\,dx, \quad y(1) = y'(1) = 0.$$

Hint: An approximate solution can be determined in the form

$$y_n(x) = (x-1)^2(a_0 + a_1 x + \ldots + a_n x^n).$$

Perform the calculation for $n = 1$.

3. Find an approximate solution of the problem of finding a minimum of the functional

$$v(y(x)) = \int_0^1 (y'^2 - y^2 - 2xy)\,dx, \quad y(0) = y(1) = 0,$$

and compare this solution with the exact solution.

Hint: An approximate solution can be determined in the form

$$y_n = x(1-x)(a_0 + a_1 x + \ldots + a_n x^n).$$

Work out this solution for $n = 0$ and $n = 1$.

4. Find an approximate solution of the problem of extrema of the functional

$$v(y(x)) = \int_1^2 \left(xy'^2 - \frac{x^2-1}{x} y^2 - 2x^2 y \right) dx, \qquad y(1) = y(2) = 0,$$

and compare this solution with the exact solution.

Hint: An approximate solution can be determined in the form

$$y = a(x-1)(x-2).$$

5. Using Ritz's method, find an approximate solution of the problem of finding minima of the functional

$$v(y(x)) = \int_0^2 (y'^2 + y^2 + 2xy) \, dx, \qquad y(0) = y(2) = 0,$$

and compare this solution with the exact solution.

Hint: Cf. Problem 3.

6. Using Ritz's method find an approximate solution of a differential equation $y'' + x^2 y = x$, $y(0) = y(1) = 0$. Find $y_2(x)$ and $y_3(x)$ and compare their values at the points $x = 0,25$, $x = 0,5$, and $x = 0,75$.

SOLUTIONS TO THE PROBLEMS

CHAPTER I

1. The extremals are the circles

$$(x-C_1)^2+y^2 = C_2^2.$$

2. The integral does not depend on the curve of integration. The variational problem does not make sense.

3. There is no extremum in the class of continuous functions.

4. The extremals are the parabolas

$$y-C_1 = \frac{1}{4}\cdot\frac{(x-C_2)^2}{C_1}.$$

5. The extremals are the rectangular hyperbolas

$$y = \frac{C_1}{x}+C_2.$$

6. $y = C_1\sin(4x-C_2)$.

7. $y = -\frac{1}{4}x^2+C_1x+C_2$.

8. $y = \sinh(C_1x+C_2)$.

9. $y = C_1e^x+C_2e^{-x}+\frac{1}{2}\sin x$.

10. $y = C_1e^{2x}+C_2e^{-2x}+C_3\cos 2x+C_4\sin 2x$.

11. $y = \dfrac{x^7}{7!}+C_1x^5+C_2x^4+C_3x^3+C_4x^2+C_5x+C_6$.

12. $y = (C_1x+C_2)\cos x+(C_3x+C_4)\sin x$, $z = 2y+y''$, and it is easy to determine z from this equation.

13. The straight lines $y = C_1x+C_2$, $z = C_3x+C_4$.

14. $\dfrac{\partial^2 z}{\partial x^2}-\dfrac{\partial^2 z}{\partial y^2} = 0$.

15. $\dfrac{\partial^2 u}{\partial x^2}+\dfrac{\partial^2 u}{\partial y^2}+\dfrac{\partial^2 u}{\partial z^2} = f(x,y,z)$.

16. $y = C_1 x^4 + C_2$.

17. $y = (C_1 x + C_2) e^x + (C_3 x + C_4) e^{-x}$.

18. $y = \frac{1}{2}\cosh x + C_1 \cos x + C_2 \sin x$.

19. $y = \frac{1}{2}x e^x + C_1 e^x + C_2 e^{-x}$.

20. $y = -\dfrac{x \cos x}{2} + C_1 \cos x + C_2 \sin x$.

CHAPTER II

1. $y = -x$ for $0 \leqslant x \leqslant 1$; $y = x - 2$ for $1 \leqslant x \leqslant 4$; and $y = x$ for $0 \leqslant x \leqslant 3$; $y = -x + 6$ for $3 \leqslant x \leqslant 4$. Along both polygonal curves the functional has a minimum.

2. Does not exist.

3. Does not exist.

4. The polygonal curves passing through the given points, consisting of linear segments with the slopes $\sqrt{3}$ and $-\sqrt{3}$.

5. $\dfrac{\varphi' - y'}{1 + y'\varphi'} = 1$, i.e. the extremals should cut the curve $y_1 = \varphi(x_1)$ along which the end point is sliding at the angle $\pi/2$.

6. $y = \dfrac{x^5}{120} + \dfrac{1}{24}(x^2 - x^3)$.

7. $y = \pm \frac{3}{4}x$ for $0 \leqslant x \leqslant \frac{16}{5}$, $y = \pm \sqrt{9 - (x - 5)^2}$ for $\frac{16}{5} \leqslant x \leqslant \frac{34}{5}$; $y = \mp \frac{3}{4}(x - 10)$ for $\frac{34}{5} \leqslant x \leqslant 10$, i.e. the curve consists of a segment of the line that is tangent to the circle, the arc of the circle, and again of the segment of the tangent to the circle.

8. $y \equiv 0$.

9. Arcs of a circle $y = \pm \sqrt{8x - x^2}$.

10. From the point A we shall go to the curve along a straight line for which $\cos(\beta - a) = v_1/v_2$, where β is the slope of this line, and a is the slope of the tangent to the curve $y = \varphi(x)$ at the point $B(x_1, y_1)$ of intersection of $y = \varphi(x)$ with a line along which the curve $y = \varphi(x)$ was approached.

Hint: This is a mixed problem of an extremum of the functional

$$\int_{x_0}^{x_1} \frac{\sqrt{1 + y'^2}}{v_1}\, dx + \int_{x_0}^{x_1} \frac{\sqrt{1 + \varphi'^2}}{v_2}\, dx.$$

CHAPTER III

1. There is a strong minimum along $y = -x^2/4 + 1$.
2. If $0 < a < \pi/4$, there is a strong minimum along $y = 0$.
3. There is no extremum in the class of continuous curves.
4. There is a strong minimum along $y = 7 - 4/x$.
5. Strong minimum along $y = 1$.
6. Strong maximum along $y = \sin 2x - 1$.
7. Strong minimum along $y = x^3$.
8. The family of extremals is $y = \sinh(C_1 x + C_2)$, C_1 and C_2 can be determined by the equations $y_0 = \sinh(C_1 x_0 + C_2)$, $y_1 = \sinh(C_1 x_1 + C_2)$. There is a weak minimum.
9. Strong minimum along $y = \frac{1}{3} e^{2x}$.
10. Strong maximum along $y = \sin 2x$.
11. Weak minimum along the line $y = \dfrac{y_1}{x_1} x$.
12. Weak minimum along the line $y = \dfrac{y_1}{x_1} x$.
13. Weak minimum along $y = x^2$.
14. Strong maximum along $y = x^3 - 1$.

CHAPTER IV

1. $y = \pm 2 \sin n\pi x$, where n is an integer.
2. $\varphi = C_1 + C_2 z$, $r = R$.
3. $y = \lambda x^2 + C_1 x + C_2$, where C_1, C_2, and λ shall be determined by the boundary conditions and by the isoperimetric condition.
4. $\dfrac{d}{dx}(p(x)y') + (\lambda r(x) - q(x))y = 0$, $y(0) = 0$, $y(x_1) = 0$.

The trivial solution $y \equiv 0$ does not satisfy the isoperimetric condition, and as it is known, non-trivial solutions exist only for some λ's, called *eigenvalues*. Therefore, λ should be an eigenvalue. One arbitrary constant of the general solution of Euler equation can be determined by the condition $y(0) = 0$, the remaining one, by the isoperimetric condition.

5. $y = -\frac{5}{2}x^2 + \frac{7}{2}x$, $z = x$.

CHAPTER V

1. $z_1 = \dfrac{5}{16a^2}(x^2 - a^2)(y^2 - b^2)$. If a better approximation is needed, we can try to find a solution in the form

$$z_2 = (x^2 - a^2)(y^2 - b^2)(a_0 + a_1(x^2 + y^2)).$$

2. $y_1 = (x-1)^2 (0.124 + 0.218x)$.

3. The exact solution is

$$y = \frac{\sin x}{\sin 1} - x.$$

4. The solution of Euler equations is $y = 3.6072 J_1(x) + 0.75195 Y_1(x) - x$, where J_1 and Y_1 are Bessel functions.

5. The exact solution is $y = \dfrac{2 \sinh x}{\sinh 2} - x$.

6. If we try to find the solution in the form

$$y_2 = x(x-1)(a_1 + a_2 x),$$

$$y_3 = x(x-1)(a_1 + a_2 x + a_3 x^2),$$

then

$$y_2 = x(x-1)(0.1708 + 0.17436x),$$

$$y_3 = x(x-1)(0.1705 + 0.1760x - 0.0018x^2).$$

At the given points the values of y_2 and y_3 are the same, up to 0.0001.

INDEX

A CATALOG OF SELECTED
DOVER BOOKS
IN SCIENCE AND MATHEMATICS

Astronomy

BURNHAM'S CELESTIAL HANDBOOK, Robert Burnham, Jr. Thorough guide to the stars beyond our solar system. Exhaustive treatment. Alphabetical by constellation: Andromeda to Cetus in Vol. 1; Chamaeleon to Orion in Vol. 2; and Pavo to Vulpecula in Vol. 3. Hundreds of illustrations. Index in Vol. 3. 2,000pp. 6⅛ x 9¼.

Vol. I: 0-486-23567-X
Vol. II: 0-486-23568-8
Vol. III: 0-486-23673-0

EXPLORING THE MOON THROUGH BINOCULARS AND SMALL TELE-SCOPES, Ernest H. Cherrington, Jr. Informative, profusely illustrated guide to locating and identifying craters, rills, seas, mountains, other lunar features. Newly revised and updated with special section of new photos. Over 100 photos and diagrams. 240pp. 8¼ x 11.   0-486-24491-1

THE EXTRATERRESTRIAL LIFE DEBATE, 1750–1900, Michael J. Crowe. First detailed, scholarly study in English of the many ideas that developed from 1750 to 1900 regarding the existence of intelligent extraterrestrial life. Examines ideas of Kant, Herschel, Voltaire, Percival Lowell, many other scientists and thinkers. 16 illustrations. 704pp. 5⅜ x 8½.   0-486-40675-X

THEORIES OF THE WORLD FROM ANTIQUITY TO THE COPERNICAN REVOLUTION, Michael J. Crowe. Newly revised edition of an accessible, enlightening book re-creates the change from an earth-centered to a sun-centered conception of the solar system. 242pp. 5⅜ x 8½.   0-486-41444-2

ARISTARCHUS OF SAMOS: The Ancient Copernicus, Sir Thomas Heath. Heath's history of astronomy ranges from Homer and Hesiod to Aristarchus and includes quotes from numerous thinkers, compilers, and scholasticists from Thales and Anaximander through Pythagoras, Plato, Aristotle, and Heraclides. 34 figures. 448pp. 5⅜ x 8½.   0-486-43886-4

A COMPLETE MANUAL OF AMATEUR ASTRONOMY: TOOLS AND TECHNIQUES FOR ASTRONOMICAL OBSERVATIONS, P. Clay Sherrod with Thomas L. Koed. Concise, highly readable book discusses: selecting, setting up and maintaining a telescope; amateur studies of the sun; lunar topography and occultations; observations of Mars, Jupiter, Saturn, the minor planets and the stars; an introduction to photoelectric photometry; more. 1981 ed. 124 figures. 25 halftones. 37 tables. 335pp. 6½ x 9¼.   0-486-42820-8

AMATEUR ASTRONOMER'S HANDBOOK, J. B. Sidgwick. Timeless, comprehensive coverage of telescopes, mirrors, lenses, mountings, telescope drives, micrometers, spectroscopes, more. 189 illustrations. 576pp. 5⅜ x 8¼. (Available in U.S. only.)   0-486-24034-7

STAR LORE: Myths, Legends, and Facts, William Tyler Olcott. Captivating retellings of the origins and histories of ancient star groups include Pegasus, Ursa Major, Pleiades, signs of the zodiac, and other constellations. "Classic."–Sky & Telescope. 58 illustrations. 544pp. 5⅜ x 8½.   0-486-43581-4

Chemistry

THE SCEPTICAL CHYMIST: THE CLASSIC 1661 TEXT, Robert Boyle. Boyle defines the term "element," asserting that all natural phenomena can be explained by the motion and organization of primary particles. 1911 ed. viii+232pp. 5⅜ x 8½.
0-486-42825-7

RADIOACTIVE SUBSTANCES, Marie Curie. Here is the celebrated scientist's doctoral thesis, the prelude to her receipt of the 1903 Nobel Prize. Curie discusses establishing atomic character of radioactivity found in compounds of uranium and thorium; extraction from pitchblende of polonium and radium; isolation of pure radium chloride; determination of atomic weight of radium; plus electric, photographic, luminous, heat, color effects of radioactivity. ii+94pp. 5⅜ x 8½.
0-486-42550-9

CHEMICAL MAGIC, Leonard A. Ford. Second Edition, Revised by E. Winston Grundmeier. Over 100 unusual stunts demonstrating cold fire, dust explosions, much more. Text explains scientific principles and stresses safety precautions. 128pp. 5⅜ x 8½.
0-486-67628-5

MOLECULAR THEORY OF CAPILLARITY, J. S. Rowlinson and B. Widom. History of surface phenomena offers critical and detailed examination and assessment of modern theories, focusing on statistical mechanics and application of results in mean-field approximation to model systems. 1989 edition. 352pp. 5⅜ x 8½.
0-486-42544-4

CHEMICAL AND CATALYTIC REACTION ENGINEERING, James J. Carberry. Designed to offer background for managing chemical reactions, this text examines behavior of chemical reactions and reactors; fluid-fluid and fluid-solid reaction systems; heterogeneous catalysis and catalytic kinetics; more. 1976 edition. 672pp. 6⅛ x 9¼.
0-486-41736-0 $31.95

ELEMENTS OF CHEMISTRY, Antoine Lavoisier. Monumental classic by founder of modern chemistry in remarkable reprint of rare 1790 Kerr translation. A must for every student of chemistry or the history of science. 539pp. 5⅜ x 8½. 0-486-64624-6

MOLECULES AND RADIATION: An Introduction to Modern Molecular Spectroscopy. Second Edition, Jeffrey I. Steinfeld. This unified treatment introduces upper-level undergraduates and graduate students to the concepts and the methods of molecular spectroscopy and applications to quantum electronics, lasers, and related optical phenomena. 1985 edition. 512pp. 5⅜ x 8½. 0-486-44152-0

A SHORT HISTORY OF CHEMISTRY, J. R. Partington. Classic exposition explores origins of chemistry, alchemy, early medical chemistry, nature of atmosphere, theory of valency, laws and structure of atomic theory, much more. 428pp. 5⅜ x 8½. (Available in U.S. only.) 0-486-65977-1

GENERAL CHEMISTRY, Linus Pauling. Revised 3rd edition of classic first-year text by Nobel laureate. Atomic and molecular structure, quantum mechanics, statistical mechanics, thermodynamics correlated with descriptive chemistry. Problems. 992pp. 5⅜ x 8½. 0-486-65622-5

ELECTRON CORRELATION IN MOLECULES, S. Wilson. This text addresses one of theoretical chemistry's central problems. Topics include molecular electronic structure, independent electron models, electron correlation, the linked diagram theorem, and related topics. 1984 edition. 304pp. 5⅜ x 8½. 0-486-45879-2

Engineering

DE RE METALLICA, Georgius Agricola. The famous Hoover translation of greatest treatise on technological chemistry, engineering, geology, mining of early modern times (1556). All 289 original woodcuts. 638pp. 6¾ x 11.   0-486-60006-8

FUNDAMENTALS OF ASTRODYNAMICS, Roger Bate et al. Modern approach developed by U.S. Air Force Academy. Designed as a first course. Problems, exercises. Numerous illustrations. 455pp. 5⅜ x 8½.   0-486-60061-0

DYNAMICS OF FLUIDS IN POROUS MEDIA, Jacob Bear. For advanced students of ground water hydrology, soil mechanics and physics, drainage and irrigation engineering and more. 335 illustrations. Exercises, with answers. 784pp. 6⅛ x 9¼.
0-486-65675-6

THEORY OF VISCOELASTICITY (SECOND EDITION), Richard M. Christensen. Complete consistent description of the linear theory of the viscoelastic behavior of materials. Problem-solving techniques discussed. 1982 edition. 29 figures. xiv+364pp. 6⅛ x 9¼.   0-486-42880-X

MECHANICS, J. P. Den Hartog. A classic introductory text or refresher. Hundreds of applications and design problems illuminate fundamentals of trusses, loaded beams and cables, etc. 334 answered problems. 462pp. 5⅜ x 8½.   0-486-60754-2

MECHANICAL VIBRATIONS, J. P. Den Hartog. Classic textbook offers lucid explanations and illustrative models, applying theories of vibrations to a variety of practical industrial engineering problems. Numerous figures. 233 problems, solutions. Appendix. Index. Preface. 436pp. 5⅜ x 8½.   0-486-64785-4

STRENGTH OF MATERIALS, J. P. Den Hartog. Full, clear treatment of basic material (tension, torsion, bending, etc.) plus advanced material on engineering methods, applications. 350 answered problems. 323pp. 5⅜ x 8½.   0-486-60755-0

A HISTORY OF MECHANICS, René Dugas. Monumental study of mechanical principles from antiquity to quantum mechanics. Contributions of ancient Greeks, Galileo, Leonardo, Kepler, Lagrange, many others. 671pp. 5⅜ x 8½. 0-486-65632-2

STABILITY THEORY AND ITS APPLICATIONS TO STRUCTURAL MECHANICS, Clive L. Dym. Self-contained text focuses on Koiter postbuckling analyses, with mathematical notions of stability of motion. Basing minimum energy principles for static stability upon dynamic concepts of stability of motion, it develops asymptotic buckling and postbuckling analyses from potential energy considerations, with applications to columns, plates, and arches. 1974 ed. 208pp. 5⅜ x 8½.
0-486-42541-X

BASIC ELECTRICITY, U.S. Bureau of Naval Personnel. Originally a training course; best nontechnical coverage. Topics include batteries, circuits, conductors, AC and DC, inductance and capacitance, generators, motors, transformers, amplifiers, etc. Many questions with answers. 349 illustrations. 1969 edition. 448pp. 6½ x 9¼.
0-486-20973-3

ROCKETS, Robert Goddard. Two of the most significant publications in the history of rocketry and jet propulsion: "A Method of Reaching Extreme Altitudes" (1919) and "Liquid Propellant Rocket Development" (1936). 128pp. 5⅜ x 8½.   0-486-42537-1

STATISTICAL MECHANICS: PRINCIPLES AND APPLICATIONS, Terrell L. Hill. Standard text covers fundamentals of statistical mechanics, applications to fluctuation theory, imperfect gases, distribution functions, more. 448pp. 5⅜ x 8½.
0-486-65390-0

ENGINEERING AND TECHNOLOGY 1650–1750: ILLUSTRATIONS AND TEXTS FROM ORIGINAL SOURCES, Martin Jensen. Highly readable text with more than 200 contemporary drawings and detailed engravings of engineering projects dealing with surveying, leveling, materials, hand tools, lifting equipment, transport and erection, piling, bailing, water supply, hydraulic engineering, and more. Among the specific projects outlined-transporting a 50-ton stone to the Louvre, erecting an obelisk, building timber locks, and dredging canals. 207pp. 8⅜ x 11¼.
0-486-42232-1

THE VARIATIONAL PRINCIPLES OF MECHANICS, Cornelius Lanczos. Graduate level coverage of calculus of variations, equations of motion, relativistic mechanics, more. First inexpensive paperbound edition of classic treatise. Index. Bibliography. 418pp. 5⅜ x 8½.   0-486-65067-7

PROTECTION OF ELECTRONIC CIRCUITS FROM OVERVOLTAGES, Ronald B. Standler. Five-part treatment presents practical rules and strategies for circuits designed to protect electronic systems from damage by transient overvoltages. 1989 ed. xxiv+434pp. 6⅛ x 9¼.   0-486-42552-5

ROTARY WING AERODYNAMICS, W. Z. Stepniewski. Clear, concise text covers aerodynamic phenomena of the rotor and offers guidelines for helicopter performance evaluation. Originally prepared for NASA. 537 figures. 640pp. 6⅛ x 9¼.
0-486-64647-5

INTRODUCTION TO SPACE DYNAMICS, William Tyrrell Thomson. Comprehensive, classic introduction to space-flight engineering for advanced undergraduate and graduate students. Includes vector algebra, kinematics, transformation of coordinates. Bibliography. Index. 352pp. 5⅜ x 8½.   0-486-65113-4

HISTORY OF STRENGTH OF MATERIALS, Stephen P. Timoshenko. Excellent historical survey of the strength of materials with many references to the theories of elasticity and structure. 245 figures. 452pp. 5⅜ x 8½.   0-486-61187-6

ANALYTICAL FRACTURE MECHANICS, David J. Unger. Self-contained text supplements standard fracture mechanics texts by focusing on analytical methods for determining crack-tip stress and strain fields. 336pp. 6⅛ x 9¼.   0-486-41737-9

STATISTICAL MECHANICS OF ELASTICITY, J. H. Weiner. Advanced, self-contained treatment illustrates general principles and elastic behavior of solids. Part 1, based on classical mechanics, studies thermoelastic behavior of crystalline and polymeric solids. Part 2, based on quantum mechanics, focuses on interatomic force laws, behavior of solids, and thermally activated processes. For students of physics and chemistry and for polymer physicists. 1983 ed. 96 figures. 496pp. 5⅜ x 8½.
0-486-42260-7

Mathematics

FUNCTIONAL ANALYSIS (Second Corrected Edition), George Bachman and Lawrence Narici. Excellent treatment of subject geared toward students with background in linear algebra, advanced calculus, physics and engineering. Text covers introduction to inner-product spaces, normed, metric spaces, and topological spaces; complete orthonormal sets, the Hahn-Banach Theorem and its consequences, and many other related subjects. 1966 ed. 544pp. 6⅛ x 9¼. 0-486-40251-7

DIFFERENTIAL MANIFOLDS, Antoni A. Kosinski. Introductory text for advanced undergraduates and graduate students presents systematic study of the topological structure of smooth manifolds, starting with elements of theory and concluding with method of surgery. 1993 edition. 288pp. 5⅜ x 8½. 0-486-46244-7

VECTOR AND TENSOR ANALYSIS WITH APPLICATIONS, A. I. Borisenko and I. E. Tarapov. Concise introduction. Worked-out problems, solutions, exercises. 257pp. 5⅝ x 8¼. 0-486-63833-2

AN INTRODUCTION TO ORDINARY DIFFERENTIAL EQUATIONS, Earl A. Coddington. A thorough and systematic first course in elementary differential equations for undergraduates in mathematics and science, with many exercises and problems (with answers). Index. 304pp. 5⅜ x 8½. 0-486-65942-9

FOURIER SERIES AND ORTHOGONAL FUNCTIONS, Harry F. Davis. An incisive text combining theory and practical example to introduce Fourier series, orthogonal functions and applications of the Fourier method to boundary-value problems. 570 exercises. Answers and notes. 416pp. 5⅜ x 8½. 0-486-65973-9

COMPUTABILITY AND UNSOLVABILITY, Martin Davis. Classic graduate-level introduction to theory of computability, usually referred to as theory of recurrent functions. New preface and appendix. 288pp. 5⅜ x 8½. 0-486-61471-9

AN INTRODUCTION TO MATHEMATICAL ANALYSIS, Robert A. Rankin. Dealing chiefly with functions of a single real variable, this text by a distinguished educator introduces limits, continuity, differentiability, integration, convergence of infinite series, double series, and infinite products. 1963 edition. 624pp. 5⅜ x 8½.
0-486-46251-X

METHODS OF NUMERICAL INTEGRATION (SECOND EDITION), Philip J. Davis and Philip Rabinowitz. Requiring only a background in calculus, this text covers approximate integration over finite and infinite intervals, error analysis, approximate integration in two or more dimensions, and automatic integration. 1984 edition. 624pp. 5⅜ x 8½. 0-486-45339-1

INTRODUCTION TO LINEAR ALGEBRA AND DIFFERENTIAL EQUATIONS, John W. Dettman. Excellent text covers complex numbers, determinants, orthonormal bases, Laplace transforms, much more. Exercises with solutions. Undergraduate level. 416pp. 5⅜ x 8½. 0-486-65191-6

RIEMANN'S ZETA FUNCTION, H. M. Edwards. Superb, high-level study of landmark 1859 publication entitled "On the Number of Primes Less Than a Given Magnitude" traces developments in mathematical theory that it inspired. xiv+315pp. 5⅜ x 8½. 0-486-41740-9

CALCULUS OF VARIATIONS WITH APPLICATIONS, George M. Ewing. Applications-oriented introduction to variational theory develops insight and promotes understanding of specialized books, research papers. Suitable for advanced undergraduate/graduate students as primary, supplementary text. 352pp. 5⅜ x 8½.
0-486-64856-7

MATHEMATICIAN'S DELIGHT, W. W. Sawyer. "Recommended with confidence" by *The Times Literary Supplement*, this lively survey was written by a renowned teacher. It starts with arithmetic and algebra, gradually proceeding to trigonometry and calculus. 1943 edition. 240pp. 5⅜ x 8½.
0-486-46240-4

ADVANCED EUCLIDEAN GEOMETRY, Roger A. Johnson. This classic text explores the geometry of the triangle and the circle, concentrating on extensions of Euclidean theory, and examining in detail many relatively recent theorems. 1929 edition. 336pp. 5⅜ x 8½.
0-486-46237-4

COUNTEREXAMPLES IN ANALYSIS, Bernard R. Gelbaum and John M. H. Olmsted. These counterexamples deal mostly with the part of analysis known as "real variables." The first half covers the real number system, and the second half encompasses higher dimensions. 1962 edition. xxiv+198pp. 5⅜ x 8½. 0-486-42875-3

CATASTROPHE THEORY FOR SCIENTISTS AND ENGINEERS, Robert Gilmore. Advanced-level treatment describes mathematics of theory grounded in the work of Poincaré, R. Thom, other mathematicians. Also important applications to problems in mathematics, physics, chemistry and engineering. 1981 edition. References. 28 tables. 397 black-and-white illustrations. xvii + 666pp. 6⅛ x 9¼.
0-486-67539-4

COMPLEX VARIABLES: Second Edition, Robert B. Ash and W. P. Novinger. Suitable for advanced undergraduates and graduate students, this newly revised treatment covers Cauchy theorem and its applications, analytic functions, and the prime number theorem. Numerous problems and solutions. 2004 edition. 224pp. 6½ x 9¼.
0-486-46250-1

NUMERICAL METHODS FOR SCIENTISTS AND ENGINEERS, Richard Hamming. Classic text stresses frequency approach in coverage of algorithms, polynomial approximation, Fourier approximation, exponential approximation, other topics. Revised and enlarged 2nd edition. 721pp. 5⅜ x 8½.
0-486-65241-6

INTRODUCTION TO NUMERICAL ANALYSIS (2nd Edition), F. B. Hildebrand. Classic, fundamental treatment covers computation, approximation, interpolation, numerical differentiation and integration, other topics. 150 new problems. 669pp. 5⅜ x 8½.
0-486-65363-3

MARKOV PROCESSES AND POTENTIAL THEORY, Robert M. Blumental and Ronald K. Getoor. This graduate-level text explores the relationship between Markov processes and potential theory in terms of excessive functions, multiplicative functionals and subprocesses, additive functionals and their potentials, and dual processes. 1968 edition. 320pp. 5⅜ x 8½.
0-486-46263-3

ABSTRACT SETS AND FINITE ORDINALS: An Introduction to the Study of Set Theory, G. B. Keene. This text unites logical and philosophical aspects of set theory in a manner intelligible to mathematicians without training in formal logic and to logicians without a mathematical background. 1961 edition. 112pp. 5⅜ x 8½.
0-486-46249-8

INTRODUCTORY REAL ANALYSIS, A.N. Kolmogorov, S. V. Fomin. Translated by Richard A. Silverman. Self-contained, evenly paced introduction to real and functional analysis. Some 350 problems. 403pp. $5\frac{3}{8}$ x $8\frac{1}{2}$. 0-486-61226-0

APPLIED ANALYSIS, Cornelius Lanczos. Classic work on analysis and design of finite processes for approximating solution of analytical problems. Algebraic equations, matrices, harmonic analysis, quadrature methods, much more. 559pp. $5\frac{3}{8}$ x $8\frac{1}{2}$. 0-486-65656-X

AN INTRODUCTION TO ALGEBRAIC STRUCTURES, Joseph Landin. Superb self-contained text covers "abstract algebra": sets and numbers, theory of groups, theory of rings, much more. Numerous well-chosen examples, exercises. 247pp. $5\frac{3}{8}$ x $8\frac{1}{2}$. 0-486-65940-2

QUALITATIVE THEORY OF DIFFERENTIAL EQUATIONS, V. V. Nemytskii and V.V. Stepanov. Classic graduate-level text by two prominent Soviet mathematicians covers classical differential equations as well as topological dynamics and ergodic theory. Bibliographies. 523pp. $5\frac{3}{8}$ x $8\frac{1}{2}$. 0-486-65954-2

THEORY OF MATRICES, Sam Perlis. Outstanding text covering rank, nonsingularity and inverses in connection with the development of canonical matrices under the relation of equivalence, and without the intervention of determinants. Includes exercises. 237pp. $5\frac{3}{8}$ x $8\frac{1}{2}$. 0-486-66810-X

INTRODUCTION TO ANALYSIS, Maxwell Rosenlicht. Unusually clear, accessible coverage of set theory, real number system, metric spaces, continuous functions, Riemann integration, multiple integrals, more. Wide range of problems. Undergraduate level. Bibliography. 254pp. $5\frac{3}{8}$ x $8\frac{1}{2}$. 0-486-65038-3

MODERN NONLINEAR EQUATIONS, Thomas L. Saaty. Emphasizes practical solution of problems; covers seven types of equations. ". . . a welcome contribution to the existing literature. . . ."–*Math Reviews*. 490pp. $5\frac{3}{8}$ x $8\frac{1}{2}$. 0-486-64232-1

MATRICES AND LINEAR ALGEBRA, Hans Schneider and George Phillip Barker. Basic textbook covers theory of matrices and its applications to systems of linear equations and related topics such as determinants, eigenvalues and differential equations. Numerous exercises. 432pp. $5\frac{3}{8}$ x $8\frac{1}{2}$. 0-486-66014-1

LINEAR ALGEBRA, Georgi E. Shilov. Determinants, linear spaces, matrix algebras, similar topics. For advanced undergraduates, graduates. Silverman translation. 387pp. $5\frac{3}{8}$ x $8\frac{1}{2}$. 0-486-63518-X

MATHEMATICAL METHODS OF GAME AND ECONOMIC THEORY: Revised Edition, Jean-Pierre Aubin. This text begins with optimization theory and convex analysis, followed by topics in game theory and mathematical economics, and concluding with an introduction to nonlinear analysis and control theory. 1982 edition. 656pp. $6\frac{1}{8}$ x $9\frac{1}{4}$. 0-486-46265-X

SET THEORY AND LOGIC, Robert R. Stoll. Lucid introduction to unified theory of mathematical concepts. Set theory and logic seen as tools for conceptual understanding of real number system. 496pp. $5\frac{5}{8}$ x $8\frac{1}{4}$. 0-486-63829-4

TENSOR CALCULUS, J.L. Synge and A. Schild. Widely used introductory text covers spaces and tensors, basic operations in Riemannian space, non-Riemannian spaces, etc. 324pp. 5⅜ x 8¼. 0-486-63612-7

ORDINARY DIFFERENTIAL EQUATIONS, Morris Tenenbaum and Harry Pollard. Exhaustive survey of ordinary differential equations for undergraduates in mathematics, engineering, science. Thorough analysis of theorems. Diagrams. Bibliography. Index. 818pp. 5⅜ x 8½. 0-486-64940-7

INTEGRAL EQUATIONS, F. G. Tricomi. Authoritative, well-written treatment of extremely useful mathematical tool with wide applications. Volterra Equations, Fredholm Equations, much more. Advanced undergraduate to graduate level. Exercises. Bibliography. 238pp. 5⅜ x 8½. 0-486-64828-1

FOURIER SERIES, Georgi P. Tolstov. Translated by Richard A. Silverman. A valuable addition to the literature on the subject, moving clearly from subject to subject and theorem to theorem. 107 problems, answers. 336pp. 5⅜ x 8½. 0-486-63317-9

INTRODUCTION TO MATHEMATICAL THINKING, Friedrich Waismann. Examinations of arithmetic, geometry, and theory of integers; rational and natural numbers; complete induction; limit and point of accumulation; remarkable curves; complex and hypercomplex numbers, more. 1959 ed. 27 figures. xii+260pp. 5⅜ x 8½. 0-486-42804-8

THE RADON TRANSFORM AND SOME OF ITS APPLICATIONS, Stanley R. Deans. Of value to mathematicians, physicists, and engineers, this excellent introduction covers both theory and applications, including a rich array of examples and literature. Revised and updated by the author. 1993 edition. 304pp. 6⅛ x 9¼. 0-486-46241-2

CALCULUS OF VARIATIONS, Robert Weinstock. Basic introduction covering isoperimetric problems, theory of elasticity, quantum mechanics, electrostatics, etc. Exercises throughout. 326pp. 5⅜ x 8½. 0-486-63069-2

THE CONTINUUM: A CRITICAL EXAMINATION OF THE FOUNDATION OF ANALYSIS, Hermann Weyl. Classic of 20th-century foundational research deals with the conceptual problem posed by the continuum. 156pp. 5⅜ x 8½. 0-486-67982-9

CHALLENGING MATHEMATICAL PROBLEMS WITH ELEMENTARY SOLUTIONS, A. M. Yaglom and I. M. Yaglom. Over 170 challenging problems on probability theory, combinatorial analysis, points and lines, topology, convex polygons, many other topics. Solutions. Total of 445pp. 5⅜ x 8½. Two-vol. set. Vol. I: 0-486-65536-9 Vol. II: 0-486-65537-7

INTRODUCTION TO PARTIAL DIFFERENTIAL EQUATIONS WITH APPLICATIONS, E. C. Zachmanoglou and Dale W. Thoe. Essentials of partial differential equations applied to common problems in engineering and the physical sciences. Problems and answers. 416pp. 5⅜ x 8½. 0-486-65251-3

STOCHASTIC PROCESSES AND FILTERING THEORY, Andrew H. Jazwinski. This unified treatment presents material previously available only in journals, and in terms accessible to engineering students. Although theory is emphasized, it discusses numerous practical applications as well. 1970 edition. 400pp. 5⅜ x 8½. 0-486-46274-9

Math–Decision Theory, Statistics, Probability

INTRODUCTION TO PROBABILITY, John E. Freund. Featured topics include permutations and factorials, probabilities and odds, frequency interpretation, mathematical expectation, decision-making, postulates of probability, rule of elimination, much more. Exercises with some solutions. Summary. 1973 edition. 247pp. 5⅜ x 8½. 0-486-67549-1

STATISTICAL AND INDUCTIVE PROBABILITIES, Hugues Leblanc. This treatment addresses a decades-old dispute among probability theorists, asserting that both statistical and inductive probabilities may be treated as sentence-theoretic measurements, and that the latter qualify as estimates of the former. 1962 edition. 160pp. 5⅜ x 8½. 0-486-44980-7

APPLIED MULTIVARIATE ANALYSIS: Using Bayesian and Frequentist Methods of Inference, Second Edition, S. James Press. This two-part treatment deals with foundations as well as models and applications. Topics include continuous multivariate distributions; regression and analysis of variance; factor analysis and latent structure analysis; and structuring multivariate populations. 1982 edition. 692pp. 5⅜ x 8½. 0-486-44236-5

LINEAR PROGRAMMING AND ECONOMIC ANALYSIS, Robert Dorfman, Paul A. Samuelson and Robert M. Solow. First comprehensive treatment of linear programming in standard economic analysis. Game theory, modern welfare economics, Leontief input-output, more. 525pp. 5⅜ x 8½. 0-486-65491-5

PROBABILITY: AN INTRODUCTION, Samuel Goldberg. Excellent basic text covers set theory, probability theory for finite sample spaces, binomial theorem, much more. 360 problems. Bibliographies. 322pp. 5⅜ x 8½. 0-486-65252-1

GAMES AND DECISIONS: INTRODUCTION AND CRITICAL SURVEY, R. Duncan Luce and Howard Raiffa. Superb nontechnical introduction to game theory, primarily applied to social sciences. Utility theory, zero-sum games, n-person games, decision-making, much more. Bibliography. 509pp. 5⅜ x 8½. 0-486-65943-7

INTRODUCTION TO THE THEORY OF GAMES, J. C. C. McKinsey. This comprehensive overview of the mathematical theory of games illustrates applications to situations involving conflicts of interest, including economic, social, political, and military contexts. Appropriate for advanced undergraduate and graduate courses; advanced calculus a prerequisite. 1952 ed. x+372pp. 5⅜ x 8½. 0-486-42811-7

FIFTY CHALLENGING PROBLEMS IN PROBABILITY WITH SOLUTIONS, Frederick Mosteller. Remarkable puzzlers, graded in difficulty, illustrate elementary and advanced aspects of probability. Detailed solutions. 88pp. 5⅜ x 8½. 0-486-65355-2

PROBABILITY THEORY: A CONCISE COURSE, Y. A. Rozanov. Highly readable, self-contained introduction covers combination of events, dependent events, Bernoulli trials, etc. 148pp. 5⅝ x 8¼. 0-486-63544-9

THE STATISTICAL ANALYSIS OF EXPERIMENTAL DATA, John Mandel. First half of book presents fundamental mathematical definitions, concepts and facts while remaining half deals with statistics primarily as an interpretive tool. Well-written text, numerous worked examples with step-by-step presentation. Includes 116 tables. 448pp. 5⅜ x 8½. 0-486-64666-1

Math–Geometry and Topology

ELEMENTARY CONCEPTS OF TOPOLOGY, Paul Alexandroff. Elegant, intuitive approach to topology from set-theoretic topology to Betti groups; how concepts of topology are useful in math and physics. 25 figures. 57pp. 5⅜ x 8½. 0-486-60747-X

A LONG WAY FROM EUCLID, Constance Reid. Lively guide by a prominent historian focuses on the role of Euclid's Elements in subsequent mathematical developments. Elementary algebra and plane geometry are sole prerequisites. 80 drawings. 1963 edition. 304pp. 5⅜ x 8½. 0-486-43613-6

EXPERIMENTS IN TOPOLOGY, Stephen Barr. Classic, lively explanation of one of the byways of mathematics. Klein bottles, Moebius strips, projective planes, map coloring, problem of the Koenigsberg bridges, much more, described with clarity and wit. 43 figures. 210pp. 5⅜ x 8½. 0-486-25933-1

THE GEOMETRY OF RENÉ DESCARTES, René Descartes. The great work founded analytical geometry. Original French text, Descartes's own diagrams, together with definitive Smith-Latham translation. 244pp. 5⅜ x 8½. 0-486-60068-8

EUCLIDEAN GEOMETRY AND TRANSFORMATIONS, Clayton W. Dodge. This introduction to Euclidean geometry emphasizes transformations, particularly isometries and similarities. Suitable for undergraduate courses, it includes numerous examples, many with detailed answers. 1972 ed. viii+296pp. 6⅛ x 9¼. 0-486-43476-1

EXCURSIONS IN GEOMETRY, C. Stanley Ogilvy. A straightedge, compass, and a little thought are all that's needed to discover the intellectual excitement of geometry. Harmonic division and Apollonian circles, inversive geometry, hexlet, Golden Section, more. 132 illustrations. 192pp. 5⅜ x 8½. 0-486-26530-7

THE THIRTEEN BOOKS OF EUCLID'S ELEMENTS, translated with introduction and commentary by Sir Thomas L. Heath. Definitive edition. Textual and linguistic notes, mathematical analysis. 2,500 years of critical commentary. Unabridged. 1,414pp. 5⅜ x 8½. Three-vol. set.
Vol. I: 0-486-60088-2   Vol. II: 0-486-60089-0   Vol. III: 0-486-60090-4

SPACE AND GEOMETRY: IN THE LIGHT OF PHYSIOLOGICAL, PSYCHOLOGICAL AND PHYSICAL INQUIRY, Ernst Mach. Three essays by an eminent philosopher and scientist explore the nature, origin, and development of our concepts of space, with a distinctness and precision suitable for undergraduate students and other readers. 1906 ed. vi+148pp. 5⅜ x 8½. 0-486-43909-7

GEOMETRY OF COMPLEX NUMBERS, Hans Schwerdtfeger. Illuminating, widely praised book on analytic geometry of circles, the Moebius transformation, and two-dimensional non-Euclidean geometries. 200pp. 5⅝ x 8¼. 0-486-63830-8

DIFFERENTIAL GEOMETRY, Heinrich W. Guggenheimer. Local differential geometry as an application of advanced calculus and linear algebra. Curvature, transformation groups, surfaces, more. Exercises. 62 figures. 378pp. 5⅜ x 8½.
0-486-63433-7

History of Math

THE WORKS OF ARCHIMEDES, Archimedes (T. L. Heath, ed.). Topics include the famous problems of the ratio of the areas of a cylinder and an inscribed sphere; the measurement of a circle; the properties of conoids, spheroids, and spirals; and the quadrature of the parabola. Informative introduction. clxxxvi+326pp. 5⅜ x 8½.
0-486-42084-1

A SHORT ACCOUNT OF THE HISTORY OF MATHEMATICS, W. W. Rouse Ball. One of clearest, most authoritative surveys from the Egyptians and Phoenicians through 19th-century figures such as Grassman, Galois, Riemann. Fourth edition. 522pp. 5⅜ x 8½.
0-486-20630-0

THE HISTORY OF THE CALCULUS AND ITS CONCEPTUAL DEVELOP-MENT, Carl B. Boyer. Origins in antiquity, medieval contributions, work of Newton, Leibniz, rigorous formulation. Treatment is verbal. 346pp. 5⅜ x 8½. 0-486-60509-4

THE HISTORICAL ROOTS OF ELEMENTARY MATHEMATICS, Lucas N. H. Bunt, Phillip S. Jones, and Jack D. Bedient. Fundamental underpinnings of modern arithmetic, algebra, geometry and number systems derived from ancient civilizations. 320pp. 5⅜ x 8½.
0-486-25563-8

THE HISTORY OF THE CALCULUS AND ITS CONCEPTUAL DEVELOP-MENT, Carl B. Boyer. Fluent description of the development of both the integral and differential calculus--its early beginnings in antiquity, medieval contributions, and a consideration of Newton and Leibniz. 368pp. 5⅜ x 8½.
0-486-60509-4

GAMES, GODS & GAMBLING: A HISTORY OF PROBABILITY AND STATISTICAL IDEAS, F. N. David. Episodes from the lives of Galileo, Fermat, Pascal, and others illustrate this fascinating account of the roots of mathematics. Features thought-provoking references to classics, archaeology, biography, poetry. 1962 edition. 304pp. 5⅜ x 8½. (Available in U.S. only.)
0-486-40023-9

OF MEN AND NUMBERS: THE STORY OF THE GREAT MATHEMATICIANS, Jane Muir. Fascinating accounts of the lives and accomplishments of history's greatest mathematical minds--Pythagoras, Descartes, Euler, Pascal, Cantor, many more. Anecdotal, illuminating. 30 diagrams. Bibliography. 256pp. 5⅜ x 8½.
0-486-28973-7

HISTORY OF MATHEMATICS, David E. Smith. Nontechnical survey from ancient Greece and Orient to late 19th century; evolution of arithmetic, geometry, trigonometry, calculating devices, algebra, the calculus. 362 illustrations. 1,355pp. 5⅜ x 8½. Two-vol. set.
Vol. I: 0-486-20429-4 Vol. II: 0-486-20430-8

A CONCISE HISTORY OF MATHEMATICS, Dirk J. Struik. The best brief history of mathematics. Stresses origins and covers every major figure from ancient Near East to 19th century. 41 illustrations. 195pp. 5⅜ x 8½.
0-486-60255-9

Physics

OPTICAL RESONANCE AND TWO-LEVEL ATOMS, L. Allen and J. H. Eberly. Clear, comprehensive introduction to basic principles behind all quantum optical resonance phenomena. 53 illustrations. Preface. Index. 256pp. 5⅜ x 8½.
0-486-65533-4

QUANTUM THEORY, David Bohm. This advanced undergraduate-level text presents the quantum theory in terms of qualitative and imaginative concepts, followed by specific applications worked out in mathematical detail. Preface. Index. 655pp. 5⅜ x 8½.
0-486-65969-0

ATOMIC PHYSICS (8th EDITION), Max Born. Nobel laureate's lucid treatment of kinetic theory of gases, elementary particles, nuclear atom, wave-corpuscles, atomic structure and spectral lines, much more. Over 40 appendices, bibliography. 495pp. 5⅜ x 8½.
0-486-65984-4

A SOPHISTICATE'S PRIMER OF RELATIVITY, P. W. Bridgman. Geared toward readers already acquainted with special relativity, this book transcends the view of theory as a working tool to answer natural questions: What is a frame of reference? What is a "law of nature"? What is the role of the "observer"? Extensive treatment, written in terms accessible to those without a scientific background. 1983 ed. xlviii+172pp. 5⅜ x 8½.
0-486-42549-5

AN INTRODUCTION TO HAMILTONIAN OPTICS, H. A. Buchdahl. Detailed account of the Hamiltonian treatment of aberration theory in geometrical optics. Many classes of optical systems defined in terms of the symmetries they possess. Problems with detailed solutions. 1970 edition. xv + 360pp. 5⅜ x 8½. 0-486-67597-1

PRIMER OF QUANTUM MECHANICS, Marvin Chester. Introductory text examines the classical quantum bead on a track: its state and representations; operator eigenvalues; harmonic oscillator and bound bead in a symmetric force field; and bead in a spherical shell. Other topics include spin, matrices, and the structure of quantum mechanics; the simplest atom; indistinguishable particles; and stationary-state perturbation theory. 1992 ed. xiv+314pp. 6⅛ x 9¼.
0-486-42878-8

LECTURES ON QUANTUM MECHANICS, Paul A. M. Dirac. Four concise, brilliant lectures on mathematical methods in quantum mechanics from Nobel Prize-winning quantum pioneer build on idea of visualizing quantum theory through the use of classical mechanics. 96pp. 5⅜ x 8½.
0-486-41713-1

THIRTY YEARS THAT SHOOK PHYSICS: THE STORY OF QUANTUM THEORY, George Gamow. Lucid, accessible introduction to influential theory of energy and matter. Careful explanations of Dirac's anti-particles, Bohr's model of the atom, much more. 12 plates. Numerous drawings. 240pp. 5⅜ x 8½. 0-486-24895-X

ELECTRONIC STRUCTURE AND THE PROPERTIES OF SOLIDS: THE PHYSICS OF THE CHEMICAL BOND, Walter A. Harrison. Innovative text offers basic understanding of the electronic structure of covalent and ionic solids, simple metals, transition metals and their compounds. Problems. 1980 edition. 582pp. 6⅛ x 9¼.
0-486-66021-4

HYDRODYNAMIC AND HYDROMAGNETIC STABILITY, S. Chandrasekhar. Lucid examination of the Rayleigh-Benard problem; clear coverage of the theory of instabilities causing convection. 704pp. 5⅜ x 8¼.   0-486-64071-X

INVESTIGATIONS ON THE THEORY OF THE BROWNIAN MOVEMENT, Albert Einstein. Five papers (1905–8) investigating dynamics of Brownian motion and evolving elementary theory. Notes by R. Fürth. 122pp. 5⅜ x 8½. 0-486-60304-0

THE PHYSICS OF WAVES, William C. Elmore and Mark A. Heald. Unique overview of classical wave theory. Acoustics, optics, electromagnetic radiation, more. Ideal as classroom text or for self-study. Problems. 477pp. 5⅜ x 8½. 0-486-64926-1

GRAVITY, George Gamow. Distinguished physicist and teacher takes reader-friendly look at three scientists whose work unlocked many of the mysteries behind the laws of physics: Galileo, Newton, and Einstein. Most of the book focuses on Newton's ideas, with a concluding chapter on post-Einsteinian speculations concerning the relationship between gravity and other physical phenomena. 160pp. 5⅜ x 8½.
0-486-42563-0

PHYSICAL PRINCIPLES OF THE QUANTUM THEORY, Werner Heisenberg. Nobel Laureate discusses quantum theory, uncertainty, wave mechanics, work of Dirac, Schroedinger, Compton, Wilson, Einstein, etc. 184pp. 5⅜ x 8½. 0-486-60113-7

ATOMIC SPECTRA AND ATOMIC STRUCTURE, Gerhard Herzberg. One of best introductions; especially for specialist in other fields. Treatment is physical rather than mathematical. 80 illustrations. 257pp. 5⅜ x 8½.   0-486-60115-3

AN INTRODUCTION TO STATISTICAL THERMODYNAMICS, Terrell L. Hill. Excellent basic text offers wide-ranging coverage of quantum statistical mechanics, systems of interacting molecules, quantum statistics, more. 523pp. 5⅜ x 8½.
0-486-65242-4

THEORETICAL PHYSICS, Georg Joos, with Ira M. Freeman. Classic overview covers essential math, mechanics, electromagnetic theory, thermodynamics, quantum mechanics, nuclear physics, other topics. First paperback edition. xxiii + 885pp. 5⅜ x 8½.   0-486-65227-0

PROBLEMS AND SOLUTIONS IN QUANTUM CHEMISTRY AND PHYSICS, Charles S. Johnson, Jr. and Lee G. Pedersen. Unusually varied problems, detailed solutions in coverage of quantum mechanics, wave mechanics, angular momentum, molecular spectroscopy, more. 280 problems plus 139 supplementary exercises. 430pp. 6½ x 9¼.   0-486-65236-X

THEORETICAL SOLID STATE PHYSICS, Vol. 1: Perfect Lattices in Equilibrium; Vol. II: Non-Equilibrium and Disorder, William Jones and Norman H. March. Monumental reference work covers fundamental theory of equilibrium properties of perfect crystalline solids, non-equilibrium properties, defects and disordered systems. Appendices. Problems. Preface. Diagrams. Index. Bibliography. Total of 1,301pp. 5⅜ x 8½. Two volumes.   Vol. I: 0-486-65015-4   Vol. II: 0-486-65016-2

WHAT IS RELATIVITY? L. D. Landau and G. B. Rumer. Written by a Nobel Prize physicist and his distinguished colleague, this compelling book explains the special theory of relativity to readers with no scientific background, using such familiar objects as trains, rulers, and clocks. 1960 ed. vi+72pp. 5⅜ x 8½.   0-486-42806-0

A TREATISE ON ELECTRICITY AND MAGNETISM, James Clerk Maxwell. Important foundation work of modern physics. Brings to final form Maxwell's theory of electromagnetism and rigorously derives his general equations of field theory. 1,084pp. 5⅜ x 8½. Two-vol. set. Vol. I: 0-486-60636-8 Vol. II: 0-486-60637-6

MATHEMATICS FOR PHYSICISTS, Philippe Dennery and Andre Krzywicki. Superb text provides math needed to understand today's more advanced topics in physics and engineering. Theory of functions of a complex variable, linear vector spaces, much more. Problems. 1967 edition. 400pp. 6½ x 9¼. 0-486-69193-4

INTRODUCTION TO QUANTUM MECHANICS WITH APPLICATIONS TO CHEMISTRY, Linus Pauling & E. Bright Wilson, Jr. Classic undergraduate text by Nobel Prize winner applies quantum mechanics to chemical and physical problems. Numerous tables and figures enhance the text. Chapter bibliographies. Appendices. Index. 468pp. 5⅜ x 8½. 0-486-64871-0

METHODS OF THERMODYNAMICS, Howard Reiss. Outstanding text focuses on physical technique of thermodynamics, typical problem areas of understanding, and significance and use of thermodynamic potential. 1965 edition. 238pp. 5⅜ x 8½. 0-486-69445-3

THE ELECTROMAGNETIC FIELD, Albert Shadowitz. Comprehensive undergraduate text covers basics of electric and magnetic fields, builds up to electromagnetic theory. Also related topics, including relativity. Over 900 problems. 768pp. 5⅝ x 8¼. 0-486-65660-8

GREAT EXPERIMENTS IN PHYSICS: FIRSTHAND ACCOUNTS FROM GALILEO TO EINSTEIN, Morris H. Shamos (ed.). 25 crucial discoveries: Newton's laws of motion, Chadwick's study of the neutron, Hertz on electromagnetic waves, more. Original accounts clearly annotated. 370pp. 5⅜ x 8½. 0-486-25346-5

EINSTEIN'S LEGACY, Julian Schwinger. A Nobel Laureate relates fascinating story of Einstein and development of relativity theory in well-illustrated, nontechnical volume. Subjects include meaning of time, paradoxes of space travel, gravity and its effect on light, non-Euclidean geometry and curving of space-time, impact of radio astronomy and space-age discoveries, and more. 189 b/w illustrations. xiv+250pp. 8⅜ x 9¼. 0-486-41974-6

THE VARIATIONAL PRINCIPLES OF MECHANICS, Cornelius Lanczos. Philosophic, less formalistic approach to analytical mechanics offers model of clear, scholarly exposition at graduate level with coverage of basics, calculus of variations, principle of virtual work, equations of motion, more. 418pp. 5⅜ x 8½. 0-486-65067-7

Paperbound unless otherwise indicated. Available at your book dealer, online at **www.doverpublications.com**, or by writing to Dept. GI, Dover Publications, Inc., 31 East 2nd Street, Mineola, NY 11501. For current price information or for free catalogues (please indicate field of interest), write to Dover Publications or log on to **www.doverpublications.com** and see every Dover book in print. Dover publishes more than 400 books each year on science, elementary and advanced mathematics, biology, music, art, literary history, social sciences, and other areas.